SpringerBriefs in Computer Science

For further volumes:
http://www.springer.com/series/10028

Yan Qiao • Shigang Chen • Tao Li

RFID as an Infrastructure

Springer

Yan Qiao
Department of Computer and Information
   Science and Engineering
University of Florida
Gainesville, USA

Shigang Chen
Department of Computer and Information
   Science and Engineering
University of Florida
Gainesville, USA

Tao Li
Department of Computer and Information
   Science and Engineering
University of Florida
Gainesville, USA

ISSN 2191-5768   ISSN 2191-5776 (electronic)
ISBN 978-1-4614-5229-4   ISBN 978-1-4614-5230-0 (eBook)
DOI 10.1007/978-1-4614-5230-0
Springer New York Heidelberg Dordrecht London

Library of Congress Control Number: 2012946730

Printed on acid-free paper

Springer is part of Springer Science+Business Media (www.springer.com)

Contents

Chapter 1
Introduction

1.1 What is RFID?

RFID (radio-frequency identification) is *the use of a wireless non-contact system that uses radio-frequency electromagnetic fields to transfer data from a tag attached to an object, for the purposes of automatic identification and tracking* [38]. The basic technologies for RFID have been around for a long time. Its root can be traced back to an espionage device designed in 1945 by Leon Theremin of the Soviet Union, which retransmitted incident radio waves modulated with audio information. After decades of development, RFID systems have gain more and more attention from both the research community and the industry.

A typical RFID system consists of one or multiple readers and numerous tags. Each tag carries a unique identifier (ID). Depending on the source of power, tags can be divided into three categories. 1) Passive tags are most widely used today. They are cheap, but do not have internal power sources. They rely on radio waves emitted from the reader for power, and have small operational ranges of a few meters. 2) Semi-passive tags contain internal batteries to power their circuits and allow longer reading distance. However, they still rely on a reader to supply its power for transmitting. 3) Active tags use their own battery power to receive and transmit information to readers. They have a much longer read range – 300 feet(91 meters) or more.

Consider a large warehouse in a distribution center of a major retailer, where hundreds of thousands of tagged commercial products are stored. In such an indoor environment, if we use passive tags, hundreds of RFID readers may have to be installed in order to access tags in the whole area, which is not only costly but also causes interference when nearby readers communicate with their tags simultaneously. It is not a good solution neither to use a mobile reader and walk through the whole area whenever we need information from tags. If the goal is to fully automate the warehouse management in a large scale, we believe battery-powered active tags are a better choice. Their much longer operational distance allows a reader to access numerous tags in a large area at one or a few fixed

Y. Qiao et al., *RFID as an Infrastructure*, SpringerBriefs in Computer Science, DOI 10.1007/978-1-4614-5230-0_1, © The Author(s) 2013

locations. Meanwhile, with richer on-tag resources, active tags are likely to gain more popularity in the future, particularly when their prices drop over time as manufactural technologies are improved and markets are expanded.

1.2   Why are RFID Technologies Important?

The barcode system brought a revolutionary change in the retail industry. Information can be embedded in the barcode. In particular, a product ID can be encoded. Once a reader retrieves the ID, it can use the ID to search a database to find all information about the product, which may include price, features, or even manufacture and shipping history. However, barcodes can only be read in close ranges with direct sight. This is fine when used for checkout in a retail store, but it is not suitable for warehouse management.

RFID technologies remove this limitation by integrating simple communication/storage/computation capacities in attachable tags, whose IDs can be read wirelessly over a distance, even when obstacles exist between tags and the RFID reader. The longer operational range makes them popular in automatic transportation payments, object tracking, and supply chain management [15, 26, 33]. Starting from August 1, 2010, Wal-Mart has begun to embed RFID tags in clothing [39]. If successful, these tags will be rolled out onto other product lines at Wal-Marts more than 3,750 U.S. stores [4]. That is one step towards cashier-less checkout, where a customer pushes his/her shopping cart to pass an RFID reader at the checkout, where information in the embedded tags is automatically read and a receipt is printed out [36].

1.3   What to be Expected Next?

In recent years, a relatively small number of research groups have been investigating novel ways in which future RFID systems can be used to solve practical problems. Of course, RFID tags may be embedded in library books, passports, driver licenses, car plates, medical products, etc. In the current application model, tags are treated as ID carriers and they are dealt with individually for the purpose of identifying the object that each tag is attached to. Now, if we make a paradigm shift from this *individual view* to a *collective view*, an array of new applications and interesting research problems will emerge. Consider a major distribution center of a large retailer, assuming it applies RFID tags to all its merchandise. These tags, which are pervasively deployed in the center, should not be treated just as ID carriers for individual objects. Collectively, they constitute a new *infrastructure*, which can be exploited for center-wide applications. If we take one step further, we can make this infrastructure more valuable by augmenting tags with miniaturized sensors, such that they report not only static ID information but also dynamic real-time information about their environment or conditions of the tags themselves. If we take another step to consider security or tag mobility, more applications and research problems open up.

1.4   Tag Estimation

Imagine a large warehouse storing thousands of refrigerators, tens of thousands of furniture pieces, or hundreds of thousands of footwear. A national retail survey showed that administration error, vendor fraud and employee theft caused about 20 billion dollars lost a year [12]. Hence, it is desirable to have a quick way of counting the number of items in the warehouse or in each section of the warehouse. To timely detect theft or management errors, such counting may be performed frequently.

If each item is attached with a RFID tag, the counting problem can be solved by a RFID reader that receives the IDs transmitted (or backscattered) from the tags [35]. However, reading the actual IDs of the tags can be time-consuming because so many of them have to be delivered in the same low-rate channel and collisions caused by simultaneous transmissions by different tags make the matter worse. Naturally, we want to design a protocol for tag estimation that minimizes the execution time.

Is time efficiency the only performance metric for the RFID estimation problem? We argue that energy cost is also an important issue that must be carefully dealt with when active tags are used to cover a large area. Active tags are powered by batteries. A longer reading range can be achieved by transmitting at higher power. Recharging batteries for tens of thousands of tags is a laborious operation, considering that the tagged products may stack up, making tags not easily accessible. To prolong the tags' lifetime and reduce the frequency of battery recharge, all functions that involve large-scale transmission by many tags should be made energy-efficient.

1.5   Sensor-augmented RFID Systems

The deployment of RFID tags will not only make the objects in a warehouse wirelessly identifiable, but also provide an "infrastructure" that we can leverage to do other things. Consider a RFID system with miniaturized sensors incorporated into tags circuit [24,26,29], enabling them to collect useful information in real time. Such system is called *sensor-augmented RFID system*. A sensor may be designed to monitor the state of the tag itself, for instance, the residual energy of the battery. In this case, the information reported to the reader can be a floating-point number reflecting the percentage of remaining energy, or simply a single bit indicating whether or not the battery needs replacement. In another example, consider a large chilled food storage facility, where each food item is attached with an RFID tag that carries a thermal sensor. An RFID reader may periodically collect temperature readings from tags to check whether any area is too hot (or too cold), which may cause food spoil (or energy waste).[1]

A sensor-augmented RFID system imposes challenges that are fundamentally different from traditional sensor networks. For example, information collection is not difficult in a classical wireless sensor network [6, 7], where each sensor implements routing/scheduling/MAC protocols. If the MAC protocol is CSMA/CA, the sensors will be able to sense the channel and transmit their information when it is idle. In addition, they are able to detect collision and use random backoff to resolve it. However, in a sensor-augmented RFID system, the simplicity of RFID tags places many constraints on the solution space, often making an otherwise easy problem difficult to solve. For example, what if the hardware of tags does not support such an MAC protocol, let alone routing/scheduling protocols? What if their simple antenna cannot sense weak signal from peers for collision avoidance, let alone performing random backoff? Hence, the challenge is to do the same work of information collection with less hardware support.

1.6 Brief Overview of State-of-the-Art

Much existing work is on designing ID-collection protocols, which read IDs from all tags in an RFID system. They mainly fall into two categories. One is *ALOHA-based* [5, 16, 30, 31, 34, 43], and the other is *tree-based* [1, 2, 25, 45]. The ALOHA-based protocols work as follows: The reader broadcasts a query request. With a certain probability, each tag chooses a time slot in the current frame to transmit its ID. If there is a collision and the reader does not acknowledge positively, the tag will continue participating in the next frame. This process repeats until all tag IDs are read successfully. Zhang et al. [42] improve the ALOHA-based protocols by extracting useful information from collision slots through analog network coding. The tree-based protocols organize all IDs in a tree of ID prefixes [1, 2, 25, 45]. Each in-tree prefix has two child nodes that have one additional bit, '0' or '1'. The tag IDs are leaves of the tree. The RFID reader walks through the tree, and requires tags with matching prefixes to transmit their IDs. Also related is a recent work that identifies tags belonging to a given set [44].

Kodialam and Nandagopal [14] estimate the number of tags in an RFID system based on the probabilistic counting methods [13]. The same authors propose a non-biased follow-up work in [15]. Han et al. [11] improve the performance of [14]. Qian et al. [27] present the Lottery-Frame scheme (LoF) for estimating the number of tags in a multiple-reader scenario. Li et al. [18] uses the maximum likelihood method. Sheng et al. design two probabilistic algorithms to identify large tag groups [32]. For the *size measurement* category, the following problems lack prior study: precisely

---

[1]If a tag reports an abnormal temperature, the reader may instruct the tag to keep transmitting beacons, which guide a mobile signal detector to locate the tag.

determining the number of tags, estimating the sizes of all groups, classifying groups based on multiple thresholds, and finding the number of new tags that enter the system and the number of existing tags that depart between two consecutive measurements. In addition, most existing work [14, 15, 27, 32] focuses on time efficiency. Their goal is to reduce the protocol execution time for solving a problem. Energy-efficient protocol design is under-studied.

Tan et al. [33] design a Trust Reader Protocol (TRP) for probabilistic missing-tag detection. Their follow-up work [31] can probabilistically identify missing tags as well as unknown tags in the system. However, it falls short of exact detection because their protocols cannot guarantee all missing tags (or unknown tags) are identified. Luo et al. [22] improves on TRP through sampling. All these protocols are designed for time efficiency, without considering how to improve energy efficiency in the detection process. Bu et al. [3] design efficient protocols to detect and pinpoint misplaced tags in a large warehouse, with the consideration of both time efficiency and energy efficiency. Luo et al. [23] reveal the energy-time tradeoff in the missing tag problem. For the *anomaly detection* category, the following problems lack prior study: exact unknown-tag detection, which is to precisely identify all unknown tags, and mixed detection of missing tags and unknown tags when both exist (probabilistic and exact versions of this problem). Although both missing tags and unknown tags are studied in [31], they are considered separately. It is unclear how their co-existence will affect each other's detection.

The idea of using RFID tags for sensing purpose has been around before [24, 29], but the problem of designing an efficient protocol to collect sensor-produced information from tags is only studied recently in [8, 41], with a primary goal of minimizing the protocol execution time. Qiao et al. [28] propose energy-efficient polling protocols for sensor-augmented RFID systems. The problem of information collection by mobile tags is not studied before.

Weis et al. [37] propose a privacy-preserving authentication protocol, in which the reader has to try all keys in the database in order to see if there exists one that produces a match with the authentication data sent from the tag. The computation overhead is prohibitively high. Yao et al. [40] use a reversible hash function, CuckooHash [10], in their authentication, which is not secure. In the weak privacy model by Lu et al. [21], a tag will keep responding the same key index to any fake readers, until it is refreshed with a new key index after a successful authentication with a legitimate reader. Hence, the key index can be used to identify the tag before refreshment. We show in [17] that all tree-based protocols [9, 19, 20, 40] cannot ensure total privacy protection, either. Therefore, the problem of privacy-preserving authentication remains open.

1.7 Outline of the Book

In Chap. 2, we discuss how to estimate the number of tags in a large RFID system. Solving the tag estimation problem incurs energy cost both at the RFID reader and at active tags. The asymmetry is that energy cost at tags should be minimized while energy cost at the reader is relatively less of a concern. We present two probabilistic algorithms that strive at saving tags' energy. The performance of the algorithms is controlled by a parameter that can be tuned to make tradeoff between energy cost and execution time.

In Chap. 3, we explain how to collect information from a sensor-augmented RFID network. We first give a lower bound on the execution time for any sensor information collection protocol. We point out that the existing ID-collection protocols are ill-fitted for this task. We then present a straightforward polling-based protocol as a baseline for comparison. Its execution time is much larger than the lower bound and its energy cost is also very high. We set forward to present more sophisticated protocols that significantly reduce the execution time toward the lower bound.

In Chap. 4, we discuss how to efficiently collect information from a subset of all tags. We first show that the standard, straightforward polling design is not energy-efficient because each tag has to continuously monitor the wireless channel and receive all tag IDs that the reader needs to collect information from, which is energy-consuming if the number of such tags is large. We show that a coded polling protocol is able to cut the amount of data each tag has to receive by half, which means that energy consumption per tag is also reduced by half. We then present two novel tag-ordering polling protocols that can reduce per-tag energy consumption by more than an order of magnitude when comparing with the coded polling protocol. In these designs, both the time efficiency and the energy efficiency are taken into consideration, whereas the tradeoff between time and energy is revealed.

References

1. Information Technology – Radio Frequency Identification for Item Management Air Interface – Part 6: Parameters for Air Interface Communications at 860-960 MHz. Final Draft International Standard ISO 18000-6 (2003)
2. Bhandari, N., Sahoo, A., Iyer, S.: Intelligent Query Tree (IQT) Protocol to Improve RFID Tag Read Efficiency. Proc. of IEEE ICIT (2006)
3. Bu, K., Xiao, B., Xiao, Q., Chen, S.: Efficient Pinpointing of Misplaced Tags in Large RFID Systems. Proc. of IEEE SECON (2011)
4. Bustillo, M.: Wal-Mart Radio Tags to Track Clothing (2010). URL http://online.wsj.com/article/SB10001424052748704421304575383213061198090.html
5. Cha, J.R., Kim, J.H.: Dynamic Framed Slotted ALOHA Algorithms using Fast Tag Estimation Method for RFID Systems. Proc. of IEEE CCNC (2006)
6. Chen, S., Fang, Y., Xia, Y.: Lexicographic Maxmin Fairness for Data Collection in Wireless Sensor Networks. IEEE Transactions on Mobile Computing 6(7), 762–776 (2007)
7. Chen, S., Yang, N.: Congestion Avoidance based on Lightweight Buffer Management in Sensor Networks. IEEE Transactions on Parallel and Distributed Systems 17(9), 934–946 (2006)
8. Chen, S., Zhang, M., Xiao, B.: Efficient Information Collection Protocols for Sensor-augmented RFID Networks. Proc. of IEEE INFOCOM (2011)
9. Dimitriou, T.: A Secure and Efficient RFID Protocol that Could Make Big Brother (partially) Obsolete. Proc. of IEEE PerCom (2006)

10. Erlingsson, Ú., Manasse, M., McSherry, F.: A Cool and Practical Alternative to Traditional Hash Tables. Proc. of WDAS (2006)
11. Han, H., Sheng, B., Tan, C.C., Li, Q., Mao, W., Lu, S.: Counting RFID Tags Efficiently and Anonymously. Proc. of IEEE INFOCOM (2010)
12. Hollinger, R., Davis, J.: National Retail Security Survey (2001). URL http://diogenesllc.com/NRSS_2001.pdf
13. Hwang, K., Vander-Zanden, B., Taylor, H.: A Linear-time Probabilistic Counting Algorithm for Database Applications. ACM Transactions on Database Systems 15(2) (1990)
14. Kodialam, M., Nandagopal, T.: Fast and Reliable Estimation Schemes in RFID Systems. Proc. of ACM MOBICOM (2006)
15. Kodialam, M., Nandagopal, T., Lau, W.: Anonymous Tracking Using RFID tags. Proc. of IEEE INFOCOM (2007)
16. Lee, S., Joo, S., Lee, C.: An Enhanced Dynamic Framed Slotted ALOHA Algorithm for RFID Tag Identification. Proc. of IEEE MOBIQUITOUS (2005)
17. Li, T., Luo, W., Mo, Z., Chen, S.: Privacy-preserving RFID Authentication based on Cryptographical Encoding. Proc. of IEEE INFOCOM (2012)
18. Li, T., Wu, S., Chen, S., Yang, M.: Energy Efficient Algorithms for the RFID Estimation Problem. Proc. of IEEE INFOCOM (2010)
19. Lu, L., Han, J., Hu, L., Liu, Y., Ni, L.: Dynamic Key-Updating: Privacy-Preserving Authentication for RFID Systems. Proc. of IEEE PerCom (2007)
20. Lu, L., Han, J., Xiao, R., Liu, Y.: ACTION: Breaking the Privacy Barrier for RFID Systems. Proc. of IEEE INFOCOM (2009)
21. Lu, L., Liu, Y., Li, X.: Refresh: Weak Privacy Model for RFID Systems. Proc. of IEEE INFOCOM (2010)
22. Luo, W., Chen, S., Li, T., Chen, S.: Efficient Missing Tag Detection in RFID Systems. Proc. of IEEE INFOCOM, mini-conference (2011)
23. Luo, W., Chen, S., Li, T., Qiao, Y.: Probabilistic Missing-tag Detection and Energy-Time Tradeoff in Large-scale RFID Systems. Proc. of ACM MobiHoc (2012)
24. Miura, M., Ito, S., Takatsuka, R., Sugihara, T., Kunifuji, S.: An Empirical Study of an RFID Mat Sensor System in a Group Home. Journal of Networks 4(2) (2009)
25. Myung, J., Lee, W.: Adaptive Splitting Protocols for RFID Tag Collision Arbitration. Proc. of ACM MOBIHOC (2006)
26. Ni, L.M., Liu, Y., Lau, Y.C., Patil, A.: LANDMARC: Indoor Location Sensing using Active RFID. ACM Wireless Networks (WINET) 10(6) (2004)
27. Qian, C., Ngan, H., Liu, Y.: Cardinality Estimation for Large-scale RFID Systems. Proc. of IEEE PerCom (2008)
28. Qiao, Y., Chen, S., Li, T., Chen, S.: Energy-efficient Polling Protocols in RFID Systems. Proc. of ACM MobiHoc (2011)
29. Ruhanen, A., Hanhikorpi, M., Bertuccelli, F., Colonna, A., Malik, W., Ranasinghe, D., Lopez, T.S., Yan, N., Tavilampi, M.: Sensor-enabled RFID Tag Handbook. BRIDGE, IST-2005-033546 (2008)
30. Sarangan, V., Devarapalli, M.R., Radhakrishnan, S.: A Framework for Fast RFID Tag Reading in Static and Mobile Environments. The International Journal of Computer and Telecommunications Networking 52(5) (2008)
31. Sheng, B., Li, Q., Mao, W.: Efficient Continuous Scanning in RFID Systems. Proc. of IEEE INFOCOM (2010)
32. Sheng, B., Tan, C.C., Li, Q., Mao, W.: Finding Popular Categories for RFID Tags. Proc. of ACM MOBIHOC (2008)
33. Tan, C., Sheng, B., Li, Q.: How to Monitor for Missing RFID Tags. Proc. of IEEE ICDCS (2008)
34. Vogt, H.: Efficient Object Identification with Passive RFID Tags. Proc. of IEEE PerCom (2002)
35. Want, R.: An Introduction to RFID Technology. Proc. of IEEE PerCom (2006)
36. Weis, S.A.: RFID (Radio Frequency Identification): Principles and Applications. MIT CSAIL (2007)

37. Weis, S.A., Sarma, S.E., Rivest, R.L., Engels, D.W.: Security and Privacy Aspects of Low-cost Radio Frequency Identification Systems. Lecture Notes in Computer Science — Security in Pervasive Computing **2802** (2004)
38. Wikipedia: Radio-frequency identification (Mar 2012). URL http://en.wikipedia.org/wiki/Radio-frequency_identification
39. Wolverton, J.: Wal-Mart to Embed RFID Tags in Clothing Beginning August 1 (2010). URL http://www.thenewamerican.com/index.php/tech-mainmenu-30/computers/4157-wal-mart-to-embed-rfid-tags-in-clothing-beginning-august-1
40. Yao, Q., Qi, Y., Han, J., Zhao, J., Li, X., Liu, Y.: Randomizing RFID Private Authentication. Proc. of IEEE PerCom (2009)
41. Yue, H., Zhang, C., Pan, M., Fang, Y., Chen, S.: A Time-efficient Information Collection Protocol for Large-scale RFID Systems. Proc. of IEEE INFOCOM (2012)
42. Zhang, M., Li, T., Chen, S., Li, B.: Using Analog Network Coding to Improve the RFID Reading Throughput. Proc. of IEEE ICDCS (2010)
43. Zhen, B., Kobayashi, M., Shimizu, M.: Framed ALOHA for Multiple RFID Objects Identification. IEICE Transactions on Communications (2005)
44. Zheng, Y., Li, M.: Fast Tag Searching Protocol for Large-Scale RFID Systems. Proc. of IEEE ICNP (2011)
45. Zhou, F., Chen, C., Jin, D., Huang, C., Min, H.: Evaluating and Optimizing Power Consumption of Anti-collision Protocols for Applications in RFID Systems. Proc. of ISLPED (2004)

Chapter 2
Tag Estimation in RFID Systems

2.1 System Model

This section introduces the tag estimation problem and the the energy issue in this problem. The communication model between RFID readers and tags is explained.

2.1.1 Tag Estimation Problem

A tag estimation problem is the problem to design efficient algorithms to estimate the number of RFID tags in a deployment area without actually reading the ID of each tag. Let N be the actual number of tags and $\hat{N}$ be the estimate. The estimation accuracy is specified by a confidence interval with two parameters: a probability value α and an error bound β, both in the range of $(0,1)$. The requirement is that the probability for $N/\hat{N}$ to fall in the interval $[1 - \beta, 1 + \beta]$ should be at least α, i.e.,

$$Prob\{(1 - \beta)\hat{N} \le N \le (1 + \beta)\hat{N}\} \ge \alpha.$$

Our goal is to reduce the energy overhead incurred to the tags during the estimation process that achieves the above accuracy.

2.1.2 Energy Issue

We consider RFID systems using active tags. Tagged goods (such as apparel) may stack in piles, and there may be obstacles, such as racks filled with merchandize, between a tag and the reader. We expect active tags are designed to transmit with significant power that is high enough to ensure reliable information delivery in such a demanding environment. Hence, energy cost due to the tags' transmissions is the main concern in our algorithm design; it increases at least in the square of the

Y. Qiao et al., *RFID as an Infrastructure*, SpringerBriefs in Computer Science,
DOI 10.1007/978-1-4614-5230-0_2, © The Author(s) 2013

maximum distance to be covered by the RFID system. Energy consumption that powers a tag's circuit for computing and receiving information is not affected by long distance and obstacles. We consider RFID systems where power consumption by tags is dominated by transmission events due to long distances that the systems need to cover. Energy consumed by the RFID reader is less of a concern. We assume the reader transmits at sufficiently high power.

2.1.3   Communication Protocol

The following communication protocol is used between a reader and tags. The reader first synchronizes the clocks of the tags and then performs a sequence of pollings. Clock synchronization only needs to happen at the beginning of the protocol execution. RFID systems operate in low-rate wireless channels. If an operation takes a short period of time, clock drift should not be a major issue in a low-rate channel.

In each polling, the reader sends out a request, which is followed by a slotted time frame during which the tags respond. The polling request from the reader carries a *contention probability* $0 < p \leq 1$ and a frame size f. Each tag will participate in the current polling with probability p. If it decides to participate, it will pick a slot uniformly at random from the frame, and transmit a bit string (called *response*) in that slot. The format of the response depends on the application. If the tag decides to not participate, it will keep silent. In our solutions, p will be set in the order of $1/N$.

If we know a lower bound N_{min} of N, the contention probability can be implemented efficiently to conserve energy. For example, a company's inventory of certain goods may be in the thousands and never before reduced below a certain number, or the company has a policy on the minimum inventory, or the RFID estimation becomes unnecessary when the number of tags is below a threshold. In these cases, we will have a lower bound N_{min}, which can be much smaller than N. If we know such a value of N_{min}, we can implement a contention probability p without requiring all tags to participate in the contention process. Since only a small number of tags actually participate in contention, energy cost is reduced. The implementation is described as follows: At the beginning of a polling, each tag makes a probabilistic decision: It goes to a standby mode for the current polling with probability $1 - 1/N_{min}$ and wakes up until the next polling starts, or it stays awake to receive the polling request with probability $1/N_{min}$ and then decides to respond with probability $min\{p \times N_{min}, 1\}$. For example, if $N = 10,000$ and $N_{min} = 1,000$, then only 10 tags stay awake in each polling.

In the above communication protocol, the reader's request may include an optional prefix and only tags that satisfy the prefix will participate in the polling. For example, suppose all tags deployed in one section of a warehouse carry the 96-bit GEN2 IDs that begin with "000" in the Serial Number field. In order to estimate the number of tags in this section, the request carries a predicate testing whether the first three bits of a tag's Serial Number is "000".

2.1.4 Empty/Singleton/Collision Slots

A slot is said to be *empty* if no tag responds (transmits) in the slot. It is called a *singleton slot* if exactly one tag responds. It is a *collision slot* if more than one tag responds. A singleton or collision slot is also called a *non-empty slot*. The Philips I-Code system [12] requires a slot length of 10 bits in order to distinguish singleton slots from collision slots. On the contrary, one bit is enough if we only need to distinguish empty slots from non-empty slots — '0' means empty and '1' means non-empty. Hence, the response will be much shorter (or consume much less energy) if an algorithm only needs to know empty/non-empty slots, instead of all three types of slots as required by [7].

In order to prolong the lifetime of tags, there are two ways to reduce their energy consumption: reducing the size of each response and reducing the number of responses. We will present algorithms that require only the knowledge of empty/non-empty slots and employ statistical methods to minimize the amount of transmission needed from the tags.

2.2 Generalized Maximum Likelihood Estimation Algorithm

The first estimator for the number of RFID tags is called the *generalized maximum likelihood estimation* (GMLE) algorithm. It fully utilizes the information from all pollings in order to minimize the number of pollings it needs to meet the accuracy requirement.

2.2.1 Overview

GMLE uses the polling protocol described in Sect. 2.1.3. The frame size f is fixed to be one slot. The RFID reader adjusts the contention probability for each polling. Let p_i be the contention probability of the ith polling. GMLE only records whether the sole slot in each polling is empty or non-empty. Based on this information, it refines the estimate $\hat{N}$ until the accuracy requirement is met. Let z_i be the slot state of the ith polling. When at least one tag responds, the slot is non-empty and $z_i = 1$. When no tag responds, it is empty and $z_i = 0$. The sequence of z_i, $i \geq 1$, forms the *response vector*.

At the ith polling, each tag has a probability p_i to transmit and, if any tag transmits, z_i will be one. Hence,

$$Prob\{z_i = 1\} = 1 - (1 - p_i)^N \approx 1 - e^{-Np_i}, \tag{2.1}$$

where N is the the actual number of tags.

If the contention probabilities of the pollings are picked too small, the response vector will contain mostly zeros. If the contention probabilities are picked too large, the response vector will contain mostly ones. Both cases do not provide sufficient statistical information for accurate estimation. As will be discussed shortly, our analysis shows that the optimal contention probability for minimizing the number of pollings is $p_i = 1.594/N$. The problem is that we do not know N (which is the quantity we want to estimate).

In order to determine p_i, GMLE consists of an *initialization phase* and an *iterative phase*. The former quickly produces a coarse estimation of N. The latter refines the contention probability and generates the estimation result.

2.2.2   Initialization Phase

We want to pick a small value for the initial contention probability p_1 at the first polling. The expected number of responding tags is Np_1. If p_1 is picked too large, a lot of tags will respond, which is wasteful because one response or many responses produce the same information — a non-empty slot. Suppose we know an upper bound N_{max} of N. This information is often available in practice. For example, we know N_{max} is 10,000 if the warehouse is designed to hold no more than 10,000 microwaves (each tagged with a RFID), or the company's inventory policy requires that in-store microwaves should not exceed 10,000, or the warehouse only has 10,000 RFID tags in use. N_{max} can be much bigger than N. We pick $p_1 = 1/N_{max}$ such that the expected number of responding tags is no more than one. If $z_1 = 0$, we multiply the contention probability by a constant $C(> 1)$, i.e., $p_2 = p_1 \times C$ for the second polling. We continue multiplying the contention probability by C after each polling until a non-empty slot is observed. When that happens (say, at the lth polling), we have a coarse estimation of N to be $1/p_l$. Then we move to the next phase. When C is relatively large, the initialization phase only takes a few pollings to complete due to the exponential increase of the contention probability.

2.2.3   Iterative Phase

This phase iteratively refines the estimation result after each polling, and terminates when the specified accuracy requirement is met. Let $\hat{N}_i$ be the estimated number of tags after the ith polling. To compute $\hat{N}_i$, the reader performs three tasks at the ith polling. First, it sets the contention probability as follows before sending out the polling request:

$$p_i = \frac{\omega}{\hat{N}_{i-1}}, \tag{2.2}$$

where $\hat{N}_{i-1}$ is the estimate after the previous polling and ω is a system parameter, which will be extensively analyzed in the next subsection. Second, based on the received z_i and the history information, the reader finds the new estimate of N that maximizes the following likelihood function:

$$L_i = \prod_{j=1}^{i}(1-p_j)^{N(1-z_j)}(1-(1-p_j)^N)^{z_j}, \qquad (2.3)$$

where $(1-p_j)^{N(1-z_j)}(1-(1-p_j)^N)^{z_j}$ is the probability for the observed state z_j of the jth polling to occur. Namely, we want to find

$$\hat{N}_i = \arg\max_{N}\{L_i\} . \qquad (2.4)$$

Third, after computing $\hat{N}_i$, the reader has to determine if the confidence interval of the new estimate meets the requirement. In the following, we show how the above tasks can be achieved.

2.2.3.1 Compute the value of $\hat{N}_i$

We compute the new estimate of N that maximizes (2.3). Since the maxima is not affected by monotone transformations, we use logarithm to turn the right side of the equation from product to summation:

$$\ln(L_i) = \sum_{j=1}^{i}\Big[N(1-z_j)\ln(1-p_j)+z_j\ln(1-(1-p_j)^N)\Big].$$

To find the maxima, we differentiate both sides:

$$\frac{\partial \ln(L_i)}{\partial N} = \sum_{j=1}^{i}\left[(1-z_j)\ln(1-p_j)-z_j\frac{(1-p_j)^N\ln(1-p_j)}{1-(1-p_j)^N}\right]. \qquad (2.5)$$

We then set the right hand side to zero and solve the equation for the new estimate $\hat{N}_i$. Note that the derivative is a monotone function of N, we can numerically obtain $\hat{N}_i$ through bisection search.

2.2.3.2 Termination Condition

Using the δ-method [2], we show that, when i is large, $\hat{N}_i$ approximately follows the Gaussian distribution:

$$Norm\left(N, \frac{(1-(1-p_i)^N)}{i(1-p_i)^N\ln^2(1-p_i)}\right).$$

The variance of $\hat{N}_i$ is

$$Var(\hat{N}_i) \approx \frac{1 - (1 - p_i)^N}{i(1 - p_i)^N \ln^2(1 - p_i)}. \tag{2.6}$$

When N is large and p_i is small, we can approximate $(1 - p_i)^N$ as e^{-Np_i} and $\ln(1 - p_i)$ as p_i. The above variance becomes

$$Var(\hat{N}_i) \approx \frac{e^{Np_i} - 1}{ip_i^2}. \tag{2.7}$$

Hence, the confidence interval of N is

$$\hat{N}_i \pm Z_\alpha \cdot \sqrt{\frac{e^{\hat{N}_i p_i} - 1}{ip_i^2}}, \tag{2.8}$$

where Z_α is the α percentile for the standard Gaussian distribution. For example, when $\alpha = 95\%$, $Z_\alpha = 1.96$. Because N is undetermined, we use $\hat{N}_i$ as an approximation when computing the standard deviation in (2.8).

The termination condition for GMLE is therefore

$$Z_\alpha \cdot \sqrt{\frac{e^{\hat{N}_i p_i} - 1}{ip_i^2}} \leq \hat{N}_i \cdot \beta, \tag{2.9}$$

where β is the error bound. The above inequality can be rewritten as

$$\sqrt{i} \geq \frac{Z_\alpha \sqrt{e^{\hat{N}_i p_i} - 1}}{\hat{N}_i p_i \beta}. \tag{2.10}$$

When i is large, the estimation changes little from one polling to the next. Hence, $p_i = \omega/\hat{N}_{i-1} \approx \omega/\hat{N}_i$. We have

$$i \geq \frac{Z_\alpha^2 \cdot (e^\omega - 1)}{\omega^2 \beta^2}. \tag{2.11}$$

Hence, if ω is determined, we can theoretically compute the approximate number of pollings that are required in order to meet the accuracy requirement. For example, if $\alpha = 95\%$, $\beta = 5\%$, and $\omega = 1.594$ (which is the optimal value to be given shortly), 2372 pollings will be required. Note that (2.11) is independent with the actual number of tags, N. Hence, our approach has perfect scalability.

Figure 2.1 shows the simulation result of GMLE when $N=10,000$, $\alpha=95\%$, $\beta = 5\%$ and $\omega = 1.594$. The simulation setup can be found in Sect. 2.4. The middle curve is the estimated number of tags, $\hat{N}_i$, with respect to the number pollings. It

Fig. 2.1 The middle curve shows the estimated number of tags with respect to the number of pollings.
The upper and lower curves show the confidence interval.

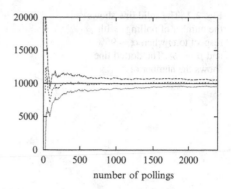

converges to the true value N represented by the central straight line. The upper and lower curves represent the 95% confidence interval, which shrinks as the number of pollings increases.

2.2.4 Determine the Value of ω

We demonstrate the impact of the value ω on two performance metrics: the *number of pollings* and the *number of tag responses* (i.e., the number of tag transmissions). The former measures the estimation time since each polling takes an equal amount of time for request/response exchange. The latter measures the energy cost because each response corresponds to one tag making one transmission in a slot.

2.2.4.1 Number of Pollings

According to (2.11), the number of pollings for meeting the accuracy requirement is $Z_\alpha^2(e^\omega - 1)/(\omega^2\beta^2)$. To find its minimum value, we differentiate it with respect to ω and let the result be zero. Solving the equation, we have $\omega = 1.594$. Hence, the optimal value of p_i that minimizes the number of pollings is

$$p_i = \frac{1.594}{\hat{N}_{i-1}}. \tag{2.12}$$

2.2.4.2 Number of Responses

We count the total number of responses during the estimation process. After a small number of pollings, the estimation will closely approximate N (see Fig. 2.1). Hence, the expected number of responses for each polling is $Np_i \approx N_{i-1}p_i = \omega$. After $Z_\alpha^2(e^\omega - 1)/(\omega^2\beta^2)$ pollings are made, the total number of responses is roughly

Fig. 2.2 The solid line shows
the number of pollings with
respect to ω when $\alpha = 95\%$
and $\beta = 5\%$. The dotted line
shows the number of
responses.

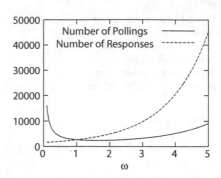

$$\frac{Z_\alpha^2 \cdot (e^\omega - 1)}{\omega^2 \beta^2} \omega = \frac{Z_\alpha^2 \cdot (e^\omega - 1)}{\omega \beta^2}. \tag{2.13}$$

Simulation results will demonstrate that the approximation in the above count is
reasonably accurate. It is an increasing function with respect to ω, which means that
a larger value of ω will lead to a larger number of responses. We give the intuition as
follows: A larger ω means a larger contention probability and thus more collisions.
Two or more responses in a collision slot produce the same amount of information
as one response in a singleton slot (see further explanation in Sect. 2.2.6). In other
words, in order to generate the necessary amount of information for meeting the
accuracy requirement, more responses must be needed if there are more collisions.

2.2.4.3 Numerical Results

in Fig. 2.2, we plot the number of pollings and the number of responses with respect
to the value of ω. The number of pollings is minimized at $\omega = 1.594$. When ω
is smaller than 1.594, its value controls the performance tradeoff between the two
metrics. When we decrease ω, the energy cost (i.e., the number of responses) drops
at the expenses of the estimation time (i.e., the number of pollings).

2.2.5 Request-less Pollings

We observe that, after a number of pollings, the value of p_i will stay in a very
small range and does not change much. It becomes unnecessary for the RFID reader
to transmit it at each polling. Hence, we can improve GMLE as follows: If the
percentage change in p_i during a certain number M_1 of consecutive pollings is
below a small threshold, the reader will broadcast a polling request, carrying the
latest value of p_i, a flag indicating that it will no longer transmit polling requests

for a certain number M_2 of slots, and the value of M_2. Without receiving further polling requests, the tags will respond with the same contention probability in the subsequent M_2 slots. This is called the *request-less pollings*. After M_2 slots, the reader will recalculate the contention probability, broadcast another polling request, carrying the new probability value, a flag, and M_2. This process repeats until the termination condition in (2.9) is met. With the threshold being 10%, $M_1 = 10$, and $M_2 = 50$, simulation results show that the performance difference caused by request-less pollings is negligibly small even though the contention probability during request-less pollings may be slightly off the value set by (2.2). Request-less pollings can also be applied to the algorithm in the next section.

2.2.6 Information Loss due to Collision

GMLE has a frame size of one slot. It obtains only binary information at each polling. No matter how many tags respond, the information that the reader receives is always the same, i.e., $z_i = 1$, which implies information loss when two or more tags decide to transmit at a polling. Let's compare two scenarios. In one scenario, only one tag responds at a polling. In the other, two tags respond. These two scenarios generate the same information but the energy cost of the second scenario is twice of the first. To address this issue, we present another algorithm that reduces the probability of collision and, moreover, compensate the impact of collision in its computation.

2.3 Enhanced Generalized Maximum Likelihood Estimation Algorithm

The *enhanced generalized maximum likelihood estimation* (EGMLE) algorithm also utilizes history information from previous pollings and uses the maximum likelihood method to estimate the number of tags. However, instead of only obtaining binary information, it computes the number of responses in each polling. Because more information can be extracted, it is able to achieve much better energy efficiency than GMLE.

2.3.1 Overview

EGMLE uses the same polling protocol as GMLE does, except that its frame size f is larger than one in order to reduce the probability of collision. The result of the ith polling, x_i, is no longer a binary value. Instead, it is an estimate of the number of tags that respond during the polling.

EGMLE takes two steps to solve the collision problem. First, it increases the frame size f such that the tags that decide to respond at a polling are likely to respond at different slots in the frame. We pick values for p_i and f such that the collision probability is very small. Second, we compensate the remaining impact of collision in our computation.

EGMLE also consists of an *initialization phase* and an *iterative phase*. The initialization phase of EGMLE is the same as the initialization phase of GMLE, except that when the RFID reader obtains the first non-zero result x_l at the lth polling with a contention probability p_l, it computes a coarse estimation of N as x_l/p_l. Then it moves to the next phase below.

2.3.2   Iterative Phase

This phase iteratively refines the estimation after each polling, and terminates when the specified accuracy requirement is met. The reader performs four tasks during the ith polling. First, it computes the contention probability before sending out the polling request.

$$p_i = \frac{\omega}{\hat{N}_{i-1}}, \tag{2.14}$$

where $\hat{N}_{i-1}$ is the estimate after the previous polling and ω is one by default. As we will show in the next subsection, performance tradeoff can be made by choosing other values for ω.

Second, the reader computes the number of responses x_i in the current frame.

Third, based on the received x_i and the history information, the reader computes the new estimate of N that maximizes the following likelihood function:

$$L_i = \prod_{j=l+1}^{i} \left[\frac{1}{\sqrt{2\pi N p_j (1 - p_j)}} \cdot e^{-\frac{((1+\varepsilon)x_j - N p_j)^2}{2N p_j (1 - p_j)}} \right], \tag{2.15}$$

where ε is introduced to compensate for collision and the iterative phase begins from the $(l+1)$th polling. The above formula and the value of ε will be derived shortly. The new estimate is

$$\hat{N}_i = \arg \max_{N} \{L_i\}. \tag{2.16}$$

Fourth, after computing $\hat{N}_i$, the reader determines if the estimate meets the accuracy requirement. In the following, we give the details of the above tasks.

Fig. 2.3 The collision probability with respect to the frame size f.

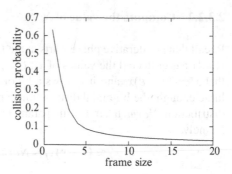

2.3.2.1 Compute the number of responses

At the ith polling, the reader measures the number of non-empty slots in the frame, denoted as x_i, which is an integer in the range of $[0..f]$. Due to possible collision, the actual number of responses, denoted as x_i^*, can be greater. Let $x_i^* = (1+\varepsilon)x_i$. The value of ε is determined below.

Since each tag independently decides to respond with probability p_i, x_i^* follows a binomial distribution, $Bino(N, p_i)$, i.e.,

$$Prob\{x_i^* = k\} = \binom{N}{k} p_i^k (1-p_i)^{N-k}. \tag{2.17}$$

Suppose ω takes the default value, 1. When i is large, N_{i-1} approximates N and thus $p_i \approx 1/N$. If N is sufficiently large, $Prob\{x_i^* = 2\} \approx 0.1839$, $Prob\{x_i^* = 3\} \approx 0.0613$, $Prob\{x_i^* = 4\} \approx 0.0153$, and the probability decreases exponentially with respect to k. $Prob\{x_i^* > 4\}$ is only about 0.0037.

Next, we compute the probability for collision to happen at the ith polling, which is denoted as $Prob_i\{collision\}$.

$$Prob_i\{collision\} = \sum_{k=2}^{N} Prob_i\{collision | x_i^* = k\} \times Prob\{x_i^* = k\}$$

$$= \sum_{k=2}^{f} (1 - \frac{P(f,k)}{f^k}) \times Prob\{x_i^* = k\} + \sum_{k=f+1}^{N} 1 \times Prob\{x_i^* = k\},$$

where $P(f,k) = f!/(f-k)!$ is the permutation function.

Figure 2.3 shows the collision probability $Prob_i\{collision\}$ with respect to f. It diminishes quickly as f increases. When $f = 10$ (which is what we use in the simulations), $Prob_i\{collision\}$ is just 0.046. With such a small probability, the chance for more than two tags involved in a collision or more than one collision at a polling is exceedingly small and thus ignored. Therefore, to approximate x_i^*, we multiply x_i by 1.046 to compensate the impact of collision. Namely, $\varepsilon = 0.046$.

2.3.2.2 Compute the value of $\hat{N}_i$

Recall that the iterative phase starts at the $(l+1)$th polling. After the ith polling, the reader has collected the values of x_j, $l < j \leq i$. By our previous analysis, we know that $x_i^* = (1+\varepsilon)x_i$ and it follows a binomial distribution $Bino(N, p_j)$. When N is large enough, the binomial distribution can be closely approximated by a Gaussian distribution $Norm(\mu_j, \sigma_j)$ with parameters $\mu_j = Np_j$ and $\sigma_j = \sqrt{Np_j(1-p_j)}$. Namely,

$$x_j^* \approx (1+\varepsilon)x_j \sim Norm(Np_j, Np_j(1-p_j)). \tag{2.18}$$

Hence, the probability for the *measured number of responses*, $(1+\varepsilon)x_j$, to occur under this distribution is $[2\pi Np_j(1-p_j)]^{-1/2}exp\{-[(1+\varepsilon)x_j - Np_j]^2/[2Np_j(1-p_j)]\}$. The likelihood function for all measured numbers of responses in the pollings, $(1+\varepsilon)x_j$, $l < j \leq i$, to occur is

$$L_i = \prod_{j=l+1}^{i} \left[\frac{1}{\sqrt{2\pi Np_j(1-p_j)}} \cdot e^{-\frac{((1+\varepsilon)x_j - Np_j)^2}{2Np_j(1-p_j)}} \right]. \tag{2.19}$$

To find the value $\hat{N}_i$ that maximizes the likelihood function, we first take logarithm on both sides of (2.19),

$$\ln(L_i) = \sum_{j=l+1}^{i} \left[\ln \frac{1}{\sqrt{2\pi Np_j(1-p_j)}} - \frac{((1+\varepsilon)x_j - Np_j)^2}{2Np_j(1-p_j)} \right]. \tag{2.20}$$

We then differentiate both sides,

$$\frac{\partial \ln(L_i)}{\partial N} = \sum_{j=l+1}^{i} \left[-\frac{1}{2N} + \frac{(1+\varepsilon)^2 x_j^2 - (Np_j)^2}{2N^2 p_j(1-p_j)} \right]$$

$$= \sum_{j=l+1}^{i} \frac{(1+\varepsilon)^2 x_j^2 - (Np_j)^2}{2N^2 p_j(1-p_j)} - \frac{i-l}{2N}. \tag{2.21}$$

Finally, we set the right side to be zero and numerically compute the value of $\hat{N}_i$.

2.3.2.3 Termination Condition

The *fisher information*[1] $\mathscr{I}(\hat{N}_i)$ of L_i is defined as follows

$$\mathscr{I}(\hat{N}_i) = -E\left[\frac{\partial^2 \ln(L_i)}{\partial N^2} \right]. \tag{2.22}$$

---

[1]The fisher information [9] is a way of measuring the amount of information that an observable random variable x carries about an unknown parameter θ upon which the likelihood function of θ, $L(\theta) = f(x; \theta)$, depends.

Fig. 2.4 The middle curve shows the estimated number of tags with respect to the number of pollings. The upper and lower curves show the confidence interval.

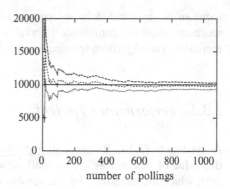

According to (2.21), we have

$$\mathscr{I}(\hat{N}_i) = E\left[\sum_{j=l+1}^{i} \frac{(1+\varepsilon)^2 x_j^2}{N^3 p_j(1-p_j)} - \frac{i-l}{2N^2}\right]$$

$$= \sum_{j=l+1}^{i} \frac{(Np_j)^2 + Np_j(1-p_j)}{N^3 p_j(1-p_j)} - \frac{i-l}{2N^2} \tag{2.23}$$

$$= \sum_{j=l+1}^{i} \frac{p_j}{N(1-p_j)} + \frac{i-l}{2N^2}. \tag{2.24}$$

Above, we have applied $E((1+\varepsilon)^2 x_j^2) = (Np_j)^2 + Np_j(1-p_j)$ in (2.23) because $(1+\varepsilon)x_j \sim Norm(Np_j, Np_j(1-p_j))$ and $E(x^2) = (E(x))^2 + Var(x)$.

Following the classical theory for MLE, when i is sufficiently large, the distribution of $\hat{N}_i$ is approximated by

$$Norm\left(N, \frac{1}{\mathscr{I}(\hat{N}_i)}\right). \tag{2.25}$$

Hence, the confidence interval is

$$\hat{N}_i \pm Z_\alpha \cdot \sqrt{\frac{1}{\mathscr{I}(\hat{N}_i)}}. \tag{2.26}$$

Note that we use $\hat{N}_i$ as an approximation for N in the computation when necessary since N is unknown. The termination condition for EGMLE to achieve the required accurary is

$$Z_\alpha \cdot \sqrt{\frac{1}{\mathscr{I}(\hat{N}_i)}} \leq \hat{N}_i \cdot \beta. \tag{2.27}$$

Figure 2.4 shows the simulation result of EGMLE when $N = 10,000$, $\alpha = 95\%$, $\beta = 5\%$, and $\omega = 1$. The middle curve is the value of $\hat{N}_i$, which converges to the value of N represented by the central straight line. The upper and lower curves represent the 95% confidence interval, which shrinks as the number of pollings increases. The algorithm terminates after 1081 pollings.

2.3.3 Performance Tradeoff

According to (2.14), the contention probability is proportional to ω. We study how the value of ω controls the tradeoff between the estimation time and the energy cost, which are measured by the number of pollings and the number of responses, respectively.

2.3.3.1 Number of Pollings

Since the MLE approach provides statistically consistent estimate, when i is large, (2.24) can be approximated as follows:

$$\mathscr{I}(\hat{N}_i) = \sum_{j=l+1}^{i} \frac{p_j}{N(1-p_j)} + \frac{i-l}{2N^2}$$

$$\approx \left(\frac{p_i}{N(1-p_i)} + \frac{1}{2N^2} \right) \cdot (i-l)$$

$$\approx \frac{2Np_i+1}{2N^2} \cdot (i-l). \tag{2.28}$$

where $p_i \ll 1$. According to (2.27), we have

$$\mathscr{I}(\hat{N}_i) \geq \left(\frac{Z_\alpha}{\hat{N}_i \cdot \beta} \right)^2 \tag{2.29}$$

(2.28) and (2.29) give us the following inequality:

$$\frac{2Np_i+1}{2N^2} \cdot (i-l) \geq \left(\frac{Z_\alpha}{\hat{N}_i \cdot \beta} \right)^2,$$

$$i \geq \frac{2Z_\alpha^2}{(2\omega+1)\beta^2}, \tag{2.30}$$

where $\hat{N}_i \approx N$ and $l \ll i$. Hence, the number of pollings it takes to achieve the accuracy requirement is $2Z_\alpha^2/[(2\omega+1)\beta^2]$.

The solid line in Fig. 2.5 shows the number of pollings with respect to ω when $\alpha = 95\%$ and $\beta = 5\%$. It is a decreasing function in ω. The reason is that a

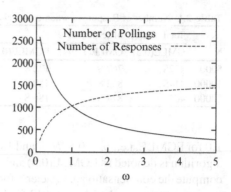

Fig. 2.5 The solid line shows the number of pollings with respect to ω when $\alpha = 95\%$ and $\beta = 5\%$. The dotted line shows the number of responses.

larger ω results in more responses (and thus more information) in each polling. Consequently, a less number of pollings is needed to achieve a certain accuracy requirement.

2.3.3.2 Number of Responses

When i is large, the expected number of responses for each polling is $Np_i \approx N_{i-1}p_i = \omega$. After $2Z_\alpha^2/[(2\omega + 1)\beta^2]$ pollings are made, the total number of responses is roughly

$$\frac{Z_\alpha^2 \cdot (e^\omega - 1)}{\omega^2 \beta^2} \omega = \frac{Z_\alpha^2 \cdot (e^\omega - 1)}{\omega \beta^2}. \tag{2.31}$$

The dotted line in Fig. 2.5 shows the number of responses with respect to ω when $\alpha = 95\%$ and $\beta = 5\%$. It is an increasing function in ω, which means that a larger value of ω will lead to a larger number of responses.

2.4 Simulations

We evaluate the performance of GMLE and EGMLE by simulations. In order to demonstrate the performance tradeoff between energy cost and estimation time, we choose two different contention probability parameters for each of the two algorithms. We use $\omega = 0.5$ and 1.594 for GMLE, i.e., $p_i = 0.5/\hat{N}_{i-1}$ and $1.594/\hat{N}_{i-1}$. Note that 1.594 is the optimal value of ω for time efficiency in GMLE. We denote the corresponding variants of the algorithm as GMLE(0.5) and GMLE(1.594).

For EGMLE, Fig. 2.5 shows that the number of pollings and the number of responses are both monotonic functions with respect to ω, which means there is no optimal ω for either energy efficiency or time efficiency. We choose $\omega = 0.5$ and

Table 2.1 Number of Responses when $\alpha = 90\%, \beta = 9\%$

| N | Total number of responses | | | | | | |
	GMLE(0.5)	GMLE(1.594)	EGMLE(0.5)	EGMLE(1.0)	UPE-O	UPE-M	EZB
5000	432S	767 S	172 S	225 S	6345 L	709 L	4342 S
10000	414S	832 S	180 S	231 S	11986 L	899 L	8683 S
20000	402S	844 S	186 S	213 S	22895 L	977 L	17366 S

1.0 for EGMLE, i.e., $p_i = 0.5/\hat{N_{i-1}}$ and $1.0/\hat{N_{i-1}}$. The corresponding variants of the algorithm is denoted as EGMLE(0.5) and EGMLE(1.0). Section 2.3.2 shows how to compute the compensation parameter ε for EGMLE(1.0), which is 0.046. Following the same steps, we obtain $\varepsilon = 0.012$ for EGMLE(0.5). We compare the algorithms with the state-of-the-art algorithms in the related work. They are the Unified Probabilistic Estimator (UPE) [7] and the Enhanced Zero-Based (EZB) estimator [8]. The original UPE, denoted as UPE-O, is very energy-inefficient because its contention probability begins from 100% and thus all tags will respond. We modify it (denoted as UPE-M) to begin from a small initial contention probability $1/N_{max}$ and keep the remaining part of UPE-O. This section shows the performance of both UPE-O and UPE-M. We run each simulation 100 times and average the outcomes.

In the initialization phase of our algorithms, let $N_{max} = 1,000,000$ and $C = 2$. The frame size in EGMLE(0.5) and EGMLE(1.0) is 10 slots. The parameters for UPE and EZB are chosen based on the original papers whenever possible. All algorithms except for UPE need only to identify empty and non-empty slots. To set a non-empty slot apart from an empty slot, a tag only needs to respond with a short bit string (one bit) to make the channel busy. UPE has to identify empty, singleton and collision slots. To set a singleton slot apart from a collision slot, many more bits (10 used by UPE) are necessary [1]. For example, CRC may be used to detect collision.

The energy cost of an algorithm depends on (1) the number of responses that all tags transmit before the algorithm terminates and (2) the size of each response. We use 'S' to mean that the response is a short bit string (in the empty/non-empty case), and 'L' to mean a long bit string (in the empty/singleton/collision case).

We do not include the simulation results for LoF [11] because its energy cost is much higher than others. Its number of responses transmitted by the tags is kN, where k is the number of frames used in the estimation process.

2.4.1 Number of Responses

The first simulation studies the number of responses in each algorithm with respect to N, α and β. Table 2.1 shows the number of responses with respect to N when $\alpha = 90\%$ and $\beta = 9\%$. GMLE and EGMLE require fewer responses than UPE and EZB. As predicted, UPE-O is energy-inefficient; UPE-M works much better.

Table 2.2 Number of Responses when $\alpha = 90\%, \beta = 6\%$

| N | Total number of responses | | | | | | |
	GMLE(0.5)	GMLE(1.594)	EGMLE(0.5)	EGMLE(1.0)	UPE-O	UPE-M	EZB
5000	1041 S	1855 S	402 S	523 S	7144 L	1811 L	7236 S
10000	1153 S	1924 S	414 S	519 S	12645 L	1687 L	14472 S
20000	1015 S	1797 S	375 S	503 S	23808 L	1814 L	28944 S

Table 2.3 Number of Responses when $\alpha = 90\%, \beta = 3\%$

| N | Total number of responses | | | | | | |
	GMLE(0.5)	GMLE(1.594)	EGMLE(0.5)	EGMLE(1.0)	UPE-O	UPE-M	EZB
5000	3927S	7341 S	1499 S	2037 S	12664 L	6426 L	27497 S
10000	3760S	7339 S	1489 S	2059 S	18023 L	6581 L	54993 S
20000	3783S	7350 S	1543 S	2002 S	28708 L	6993 L	109987 S

Table 2.4 Number of Responses when $\alpha = 95\%, \beta = 9\%$

| N | Total number of responses | | | | | | |
	GMLE(0.5)	GMLE(1.594)	EGMLE(0.5)	EGMLE(1.0)	UPE-O	UPE-M	EZB
5000	603S	1112 S	258 S	330 S	6715 L	1073 L	4342 S
10000	669S	1120 S	247 S	304 S	12062 L	961 L	8683 S
20000	680S	1197 S	262 S	320 S	23345 L	1136 L	17366 S

Table 2.5 Number of Responses when $\alpha = 95\%, \beta = 6\%$

| N | Total number of responses | | | | | | |
	GMLE(0.5)	GMLE(1.594)	EGMLE(0.5)	EGMLE(1.0)	UPE-O	UPE-M	EZB
5000	1340 S	2515 S	581 S	736 S	7712 L	2598 L	10130 S
10000	1354 S	2511 S	596 S	736 S	13477 L	2318 L	20261 S
20000	1381 S	2630 S	555 S	749 S	24631 L	2510 L	40521 S

The best algorithm is EGMLE(0.5), whose number of responses is about one fifth of what UPE-M requires and one ninetieth of what EZB requires when N is 20,000. Moreover, each response in UPE is much longer.

GMLE(0.5) has a smaller energy cost than GMLE(1.594). For example, $N = 10,000$, the ratio between the number of responses by GMLE(1.594) and that by GMLE(0.5) is 2.01, which is close to the theoretically-computed ratio of 1.90 in Fig. 2.2. Similarly, EGMLE(0.5) is more energy efficient than EGMLE(1.0). When $N = 10,000$, the ratio between the number of responses by GMLE(1.594) and that by GMLE(0.5) is 1.28, which is also close to the theoretical value of 1.34 in Fig. 2.5.

We vary α from 90% to 95% and to 99%, and vary β from 9% to 6% and to 3%. Tables 2.2 to 2.9 show similar comparison under different values of α and β values. In all cases, the number of responses increases when α increases or β decreases, and except for EZB, the number does not vary much with respect to N, meaning that all algorithms except for EZB achieve good scalability. The ratio between the numbers for different algorithms appears to be quite stable under different parameter settings.

Table 2.6 Number of Responses when $\alpha = 95\%, \beta = 3\%$

	Total number of responses						
N	GMLE(0.5)	GMLE(1.594)	EGMLE(0.5)	EGMLE(1.0)	UPE-O	UPE-M	EZB
5000	5687 S	10493 S	2181 S	2915 S	14678 L	8858 L	39074 S
10000	5673 S	10286 S	2267 S	2924 S	20845 L	9364 L	78148 S
20000	5588 S	10637 S	2217 S	2990 S	32339 L	9683 L	156297 S

Table 2.7 Number of Responses when $\alpha = 99\%, \beta = 9\%$

	Total number of responses						
N	GMLE(0.5)	GMLE(1.594)	EGMLE(0.5)	EGMLE(1.0)	UPE-O	UPE-M	EZB
5000	1040 S	2162 S	427 S	453 S	7240 L	1726 L	7236 S
10000	1071 S	2135 S	416 S	529 S	12842 L	1906 L	14472 S
20000	1017 S	1916 S	439 S	573 S	23982 L	1819 L	28944 S

Table 2.8 Number of Responses when $\alpha = 99\%, \beta = 6\%$

	Total number of responses						
N	GMLE(0.5)	GMLE(1.594)	EGMLE(0.5)	EGMLE(1.0)	UPE-O	UPE-M	EZB
5000	2527 S	4785 S	965 S	1269 S	9679 L	4311 L	17366 S
10000	2527 S	4637 S	973 S	1248 S	15336 L	4130 L	34733 S
20000	2440 S	4580 S	991 S	1293 S	26128 L	4044 L	69465 S

Table 2.9 Number of Responses when $\alpha = 99\%, \beta = 3\%$

	Total number of responses						
N	GMLE(0.5)	GMLE(1.594)	EGMLE(0.5)	EGMLE(1.0)	UPE-O	UPE-M	EZB
5000	9693 S	18690 S	3818 S	4993 S	21823 L	16705 L	65124 S
10000	9606 S	18223 S	3791 S	4998 S	27667 L	15882 L	130247 S
20000	9385 S	17735 S	3847 S	5027 S	38935 L	16471 L	260495 S

2.4.2   Total Number of Bits Transmitted

The second simulation evaluates the energy cost of the algorithms. As mentioned before, one bit is enough to separate empty/non-empty slot. Hence, the response of GMLE, EGMLE and EZB is one bit long. A response in UPE-M is 10 bits long [7]. We compare the total number of bits transmitted by all tags before each algorithm terminates. We omit the results for UPE-O, which are much worse than the results of UPE-M. Figure 2.6 shows the simulation results with respect to N when $\alpha = 90\%, \beta = 9\%, 6\%$ and 3%. For example, when $\alpha = 90\%$, $\beta = 3\%$, and $N = 20,000$, the ratio between the number of bits transmitted by UPE-M (EZB) and

Fig. 2.6 Numbers of bits
transmitted when
$\alpha = 90\%, \beta = 9\%, 6\%$
and 3%.

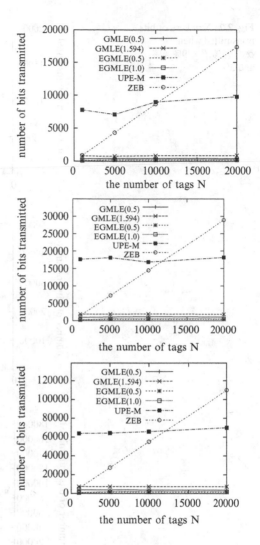

that by our best estimator EGMLE(0.5) is 45.32 (71.28). Figures 2.7 and 2.8 show
the comparison under different β values when $\alpha = 95\%$ and 99%, respectively.
Their results are similar to Fig. 2.6. It should be noted that the number of bits
transmitted is not an accurate measurement of the energy cost because it ignores
the energy spent to power up the radio and synchronize with the reader. However,
combining the number of bits and the number of transmissions (in the previous
subsection) still gives a good idea on how energy-efficient each algorithm is.

Fig. 2.7 Numbers of bits
transmitted when
$\alpha = 95\%, \beta = 9\%, 6\%$
and 3%.

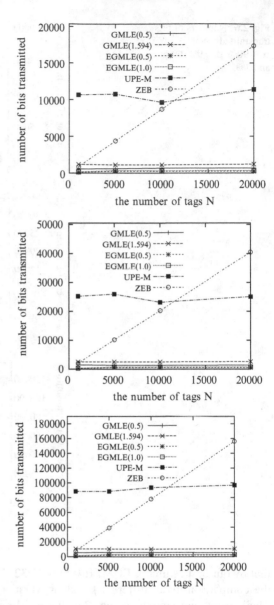

2.4.3   Estimation Time

The third simulation compares the time it takes for each algorithm to complete the
estimation of N. Based on the specification of the Philips I-Code system [12], after
the required waiting times (e.g., gap between transmissions) are included, it can
be calculated that a RFID reader needs 0.4 ms to detect an empty slot, 0.8 ms
to detect a collision or a singleton slot, and 1 ms to broadcast a polling request.
Hence, GMLE, EGMLE and EZB requires a slot length of 0.4 ms, while UPE-M

Fig. 2.8 Numbers of bits transmitted when $\alpha = 99\%, \beta = 9\%, 6\%$ and 3%.

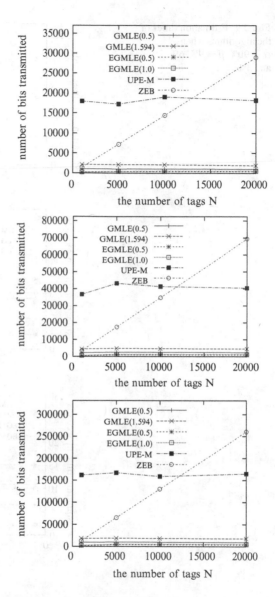

requires a slot length of 0.8 *ms*. Recall that the contention probability takes the form of $\omega/\hat{N}_i$, where ω is a known constant. Thus the reader transmits $\hat{N}_i$ instead of the actual probability value in the polling requests. If we assume N_{max} is no more than a million, then 20 bits for $\hat{N}_i$ are sufficient. GMLE has a fixed frame size of one slot. EGMLE has a fixed frame size of 10 slots. EZB and UPE-M also have pre-determined frame sizes. Let $\alpha = 90\%, \beta = 9\%, 6\%$ and 3%. The three plots in Fig. 2.9 show the estimation times of the algorithms with respect to the number of tags in the deployment. The times grow very slowly as the number of tags increase, which suggests the algorithms all scale well. In the first plot of Fig. 2.9, UPE-M

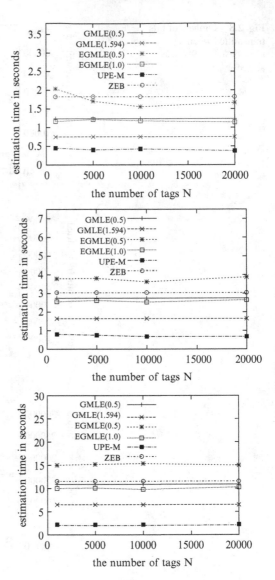

Fig. 2.9 Estimation times of the algorithms when $\alpha = 90\%, \beta = 9\%, 6\%$ and 3%.

takes the least amount of time, only about 0.5 second, to estimate 20,000 tags, while the other algorithms take between 0.7 to 2.0 seconds. GMLE(1.594) takes less estimation time than GMLE(0.5) and the ratio is 0.61, which is consistent with the theoretical value of 0.58 in Fig. 2.2. Similarly, EGMLE(1.0) takes less time than EGMLE(0.5) and the ratio is 0.68, which is also consistent with the theoretical value of 0.67 in Fig. 2.5. Figures 2.10 and 2.11 show similar simulation results when $\alpha = 95\%$ and 99%, respectively. Even though the new algorithms take longer to complete, their estimation time is still small. We believe the extra time needed can be well justified for the large energy saving.

Fig. 2.10 Estimation times
of the algorithms when
$\alpha = 95\%, \beta = 9\%, 6\%$
and 3%.

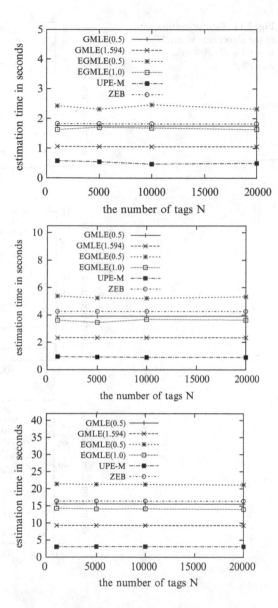

There exists a performance tradeoff between GMLE and EGMLE. In the
previous two subsections, we have examined energy cost in terms of number of
responses and number of transmitted bits. EGMLE always performs better than
GMLE. In this subsection, we compare estimation time of our two methods. GMLE
performs better than EGMLE. Because the focus of this work is on energy efficiency,
we regard EGMLE as the best estimator for energy saving.

Fig. 2.11 Estimation times
of the algorithms when
$\alpha = 99\%, \beta = 9\%, 6\%$
and 3%.

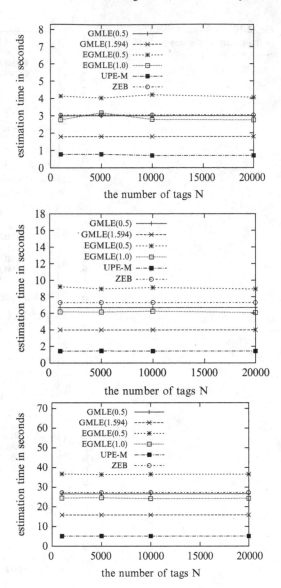

2.5 Other Methods

Instead of identifying individual RFID tags, Floerkemeier [4, 5] studies the problem
of estimating the cardinality of a tag set based on the number of empty slots.
The proposed scheme employs a Bayesian probability estimation to achieve fast
estimation. The scheme is similar to hash-based estimators [3, 14] and the difference
is discussed in [8]. In Kodialam and Nandagopal's approach [7], information from
tags are collected by a RFID reader in a series of time frames. Each frame consists

of a number of slots, and the tags probabilistically respond in those slots. Using the probabilistic counting methods, the reader estimates the number of tags based on the number of empty slots or the number of collision slots in each frame. Their best estimator is called the Unified Probabilistic Estimator (UPE). A follow-up work by the same authors proposes the Enhanced Zero-Based Estimator (EZB) [8], which makes its estimation based on the number of empty slots. The focus of the above estimators is to reduce the time it takes a reader to complete the estimation process. Because their goal is not conserving energy for active tags, their design is not geared towards reducing the number of transmissions made by the tags.

The Lottery-Frame scheme (LoF) [11] by Qian et al. employs a geometric distribution-based scheme to determine which slot in a time frame each tag will respond. It significantly reduces the estimation time when comparing with UPE. However, every tag must respond in each of the time frames, resulting in large energy cost when active tags use their own power to transmit. The First Non-Empty slots Based algorithm (FNEB) [6] uses the slot number of the first reply from tags in a frame to count RFID tags in both static and dynamic environments.

Also related is a novel security protocol proposed by Tan et al. to monitor the event of missing tags in the presence of dishonest RFID readers [13]. In order to prevent a dishonest reader from replaying previously collected information, they maintain a timer in the server and periodically update the system clock. Li et al. [10] design a series of efficient protocols that employ novel techniques to identify missing tags in large-scale RFID systems.

2.6 Summary

This chapter presents two probabilistic algorithms for estimating the number of RFID tags in a region. Solving the tag estimation problem incurs energy cost both at the RFID reader and at active tags. The asymmetry is that energy cost at tags should be minimized while energy cost at the reader is relatively less of a concern because the reader's battery can be replaced easily or it may be powered by an external source. To exploit this asymmetry, the probabilistic algorithms trade more energy cost at the reader for less cost at the tags. The performance of the algorithms is controlled by a parameter ω, specifying the contention probability that tags use to decide whether they will transmit. By modifying this parameter, the algorithms can make tradeoff between energy cost and estimation time.

References

1. EPC Radio-Frequency Identity Protocols Class-1 Generation-2 UHF RFID Protocol for Communications at 860 MHz - 960 MHz Version 1.0.9. http://www.epcglobalinc.org/standards/uhfc1g2/uhfc1g2_1_0_9-standard-20050126.pdf (2005)
2. Casella, G., Berger, R.L.: Statistical Inference. 2nd edition, Duxbury Press (2002)

3. Durand, M., Flajolet, P.: LogLog Counting of Large Cardinalities. Proc. of European Symposium on Algorithms (2003)
4. Floerkemeier, C.: Transmission Control Scheme for Fast RFID Object Identification. IEEE Percom Workshop on Pervasive Wireless Networking (2006)
5. Floerkemeier, C., Wille, M.: Comparison of Transmission Schemes for Framed ALOHA based RFID Protocols. Workshop on RFID and Extended Network Deployment of Technologies and Applications (2006)
6. Han, H., Sheng, B., Tan, C.C., Li, Q., Mao, W., Lu, S.: Counting RFID Tags Efficiently and Anonymously. Proc. of IEEE INFOCOM (2010)
7. Kodialam, M., Nandagopal, T.: Fast and Reliable Estimation Schemes in RFID Systems. Proc. of ACM MOBICOM (2006)
8. Kodialam, M., Nandagopal, T., Lau, W.: Anonymous Tracking using RFID tags. Proc. of IEEE INFOCOM (2007)
9. Lehmann, Casella, G.: Theory of Point Estimation. Springer, 2nd edition (1998)
10. Li, T., Chen, S., Ling, Y.: Identifying the Missing Tags in a Large RFID System. Proc. of ACM Mobihoc (2010)
11. Qian, C., Ngan, H., Liu, Y.: Cardinality Estimation for Large-scale RFID Systems. Proc. of IEEE PerCom (2008)
12. Semiconductors, P.: I-CODE Smart Label RFID Tags. http://www.nxp.com/documents/data_sheet/SL092030.pdf (2004)
13. Tan, C., Sheng, B., Li, Q.: How to Monitor for Missing RFID Tags. Proc. of IEEE ICDCS (2008)
14. Whang, K., Vander-Zanden, B., Taylor, H.: A Linear Time Probabilistic Counting Algorithm for Database Applications. ACM Transactions on Database Systems (1990)

Chapter 3
Collecting Information from Sensor-augmented RFID Systems

3.1 System Model

In this section, we first introduces the information collection problem in sensor-augmented RFID systems and make assumptions. Then a theoretical lower bound on the execution time is provided as a basis to evaluate the protocols.

3.1.1 Problem

Consider a RFID system with a large number of active tags deployed in a region. Each tag is equipped with a sensor that generates a certain type of information, which can be one bit or multiple bits. In the rest of this chapter, we will refer to a tag's sensor information simply as a tag's information. We assume that the RFID reader and the tags transmit with sufficient power such that they can communicate over a long distance. Communications between the reader and the tags are time-slotted. The reader's signal will synchronize the clocks of the tags. Generally speaking, communications are driven by the reader in a request-and-response pattern. The reader issues a request, which is followed by a tag's response or a slotted time frame in which multiple tags respond.

The problem is to design a protocol for a reader to periodically collect information from the tags. Our goal is to minimize the execution time of the information collection protocol, so that it uses as little time as possible to gather data from the tags.

3.1.2 Assumption

We assume that the RFID reader has access to a database that stores the IDs of all tags. This is a reasonable assumption for RFID-assisted warehouse management, where the tag IDs are read into a database when new objects are moved into the system and they are removed from the database when the objects are moved out. Even if this operation is not performed, there are many protocols that are designed to collect all tag IDs from the system (see the introduction). Once the tag IDs are collected, we can use the protocols in this chapter to periodically collect information from the tags.

The set of tags in a warehouse changes over time. If the execution time of the protocols is short, the set of tags is likely to be stable during the protocol execution. However, even if the set of tags changes, the reader can simply ignore the tags that are added or removed from the system during the protocol execution. The reader will start to collect information from new tags in the next execution of the protocol.

Actions may need to be taken after the sensor information indicates a problem: For example, the battery of a sensor needs to be replaced or the temperature in a certain section of a chilled storage is too high. The problem of physically locating the alarm-raising tag is beyond the scope of this book. One possible method is to instruct the tag to keep transmitting so that it can be located by a mobile device that detects the direction and distance of a transmitting target. Another approach is to use a localization protocol [6].

3.1.3   Performance Lower Bound and ID-collection Protocols

Let t_{id} be the length of a time slot that the reader uses to broadcast a tag ID, which is 96 bits in the Gen2 standard. Note that the amount of time it takes the reader to transmit an ID may be different from what it takes a tag to transmit an ID because the reader and the tags may operate at different transmission rates [9]. Let t_{inf} be the length of a time slot for a tag to transmit its information. The value of t_{inf} depends on the number of bits that the information contains, which is application-specific. Let n be the number of tags in the system. A lower bound for any protocol to collect information from all tags is $n \times t_{inf}$, which is the aggregate time for all tags to report their information. This lower bound is not achievable because it takes additional time for the reader to send its request(s). However, we can design a protocol whose expected execution time is reasonably close to this lower bound.

Collecting sensor information from tags is a different problem than collecting IDs from the tags. In fact, solutions to these two problems are complementary in practice. First, the ID of a tag only needs to be read once when the tag enters the system and it is removed when the tag exits. Sensor information needs to be collected periodically. Second, tag IDs are a set of numbers, whereas sensor information is not only a set of sensor readings but also a mapping from the readings to the tags where each sensor reading takes place.

One may argue that an ID-collection protocol can piggyback a tag's sensor information when it reads the tag's ID. There are two major types of ID-collection protocols: ALOHA-based or tree-based. It is well known that, for any ALOHA-based protocol [3, 4, 8, 10–12], the optimal execution time for reading n tags is $e \times n \times T$ [7], where e is the natural constant and T is the length of a time slot in which a tag's ID and its sensor information can be transmitted.[1] Note that $n \times T$ is not achievable due to collision in ALOHA. For tree-based protocols [1,2,5,13], analytical and simulation results have shown that their best performance is comparable to the best of the ALOHA-based protocols.

We know that a lower bound for only collecting sensor information from n tags is $n \times t_{inf}$, where t_{inf} can be as little as one seventh of T when one bit information is reported (see Sect. 3.6). Hence, the optimal execution time $e \times n \times T$ of an ID-collection protocol can be almost twenty times of our lower bound. In contrast, the best protocol specifically designed for information collection achieves an execution time within 1.44 times the lower bound, as we will see later. The reason is that when we periodically collect sensor information from tags, the IDs of the tags are supposed to be already known and in fact the protocol design relies on the knowledge of these IDs to help avoid radio collisions in order to improve time efficiency.

3.2 Polling-based Information Collection Protocol

The baseline protocol is called the *polling-based information collection protocol* (PIC). It is very simple. The RFID reader broadcasts the tag IDs one after another. After it transmits an ID, it waits for a period of t_{inf} to receive the information of the corresponding tag. Hence, the time to collect information from one tag is $t_{id} + t_{inf}$. The total execution time of the protocol for collecting information from all tags is $n \times (t_{id} + t_{inf})$.

PIC has two major limitations. First, its execution time is much larger than the optimal value $n \times t_{inf}$. Using the parameters in [9], we find that t_{id} can be twelve times of t_{inf} when the information that a tag reports is one bit. Hence, $n \times (t_{id} + t_{inf})$ is up to thirteen times of the lower bound, which leaves much room for improvement. Second, each tag must continuously listen to the communication channel until its ID is received. If battery-powered active tags are used, this will cause significant energy overhead because each tag has to keep powering its circuit and may have to receive thousands of tag IDs before finding its own. In the following, we present two protocols that solve the energy problem and are much more time-efficient.

---

[1]In an ALOHA-based protocol, we cannot let tags only transmit their sensor readings without sending their IDs because we need to know which tag each senor reading belongs to.

3.3   Single-hash Information Collection Protocol

In this section, we present a Single-hash Information Collection protocol (SIC) that avoids the transmission of tag IDs and does not require the tags to continuously listen to the channel.

3.3.1   Protocol Overview

The execution of the SIC protocol consists of multiple phases. Every phase has the same structure: It begins with *an information collection request* sent from the reader to the tags, followed by a *slotted time frame*, in which some tags are scheduled to transmit their information. As we will explain later, about 36.8% of the tags are scheduled for transmission in the first phase, 36.8% of the remaining tags are scheduled in the second phase, ..., until all tags are scheduled for transmission. We stress that each tag will be scheduled only once in one of the phases, and it will be *assigned* by the reader to a unique slot in the time frame of that phase.

Next, we will first explain how to assign tags to slots, and then give the protocol details.

3.3.2   Assigning Tags to Time Slots Using a Hash Function

Consider an arbitrary phase. Suppose there are n' tags from which the reader has not yet received information. Only these tags are considered for slot assignment because the information of other tags has been received in the previous phases. Clearly, $n' = n$ for the first phase.

The reader always sets the number of slots in the frame equal to the number of tags it considers for slot assignment. Namely, the frame size is n'. Before the reader sends out a request, it has to determine which tags should transmit in this phase and which slots in the frame they should be assigned to. To avoid collision, it should never assign more than one tag to a slot. Because the number of tags is equal to the number of slots, it is not difficult for the reader to construct a one-to-one mapping from the tags to the slots. But it is too costly to inform the tags about this mapping, especially when the set of tags may change each time the protocol is executed.

One solution is for the reader to use a hash function H to map the tags to the slots, while the tags use the same hash function to determine which slots they should use. The hash function takes the ID of a tag and a random number r as input and produces a pseudo random number $H(ID, r)$ as output, which is used as the slot index that the tag is mapped to. However, this approach does not ensure one-to-one mapping. Multiple tags may be mapped to the same slot. In this case, these tags cannot transmit in the slot because otherwise collision would occur. The slot is thus *wasted*. If no tag is mapped to a slot, that slot is also *wasted*. Only when one and only one tag is mapped to a slot, the reader will assign the tag to the slot. In this case, we say the slot is *useful*. How to inform tags which slots are useful so that the tags assigned to them will transmit in these slots? We introduce an indicator vector, which is described in the next subsection.

3.3.3   Protocol Description

SIC consists of multiple phases. In each phase, the reader sends out a request and then tags transmit their information in the subsequent frame. Before sending out the request, the RFID reader has to assign tags to the slots of the frame. It picks a random number r and uses the hash function to map the IDs of the tags to the slots. After determining which slots are useful and which have to be wasted, the reader constructs an n'-bit *indicator vector*, where n' is the number of tags that are considered for slot assignment. Recall that it is also the number of slots in the frame. Each bit in the vector corresponds to a slot in the frame at the same index location. If the slot is useful (i.e., one and only one tag is mapped to it), the bit value is 1; otherwise, it is 0.

The request broadcast by the reader consists of the information type to be reported, the frame size (i.e., number of slots in the frame), a random number r, and the indicator vector, where r is used by the hash function and it is different in each phase. If the vector is too long, the reader divides it into segments of 96 bits (equivalent to the length of a tag ID) and transmits each segment in a time slot of length t_{id}.

Using the same hash function, a tag knows the index i of the slot it is mapped to. After the tag receives the request, it knows whether its slot will be useful or not by examining the ith bit in the indicator vector. If the bit is 0, the tag will not transmit. If the bit is 1, the tag will transmit its information during the ith slot in the frame and it will not participate in the remaining phases.

It should be noted that the tag does not have to receive the whole indicator vector. It knows the index i of the bit it looks for. Hence, it also knows which segment of the indicator vector it must receive. The tag can be in a stand-by mode to conserve energy at times other than when it receives its segment of the indicator vector or transmits its information.

The first phase considers n tags for slot assignment and its frame has n slots. Each subsequent phase considers a fewer number of tags and has a smaller frame accordingly. The protocol terminates after all tags report their information. Alternatively, the reader may stop the SIC protocol when the number of remaining tags is fewer than a small threshold, and then it invokes the PIC protocol to collect information from these tags.

3.3.4 Expected Execution Time

We derive the expected execution time of the SIC protocol. Consider an arbitrary tag x and an arbitrary phase that x participates. Let n' be the number of tags that are considered for slot assignment in this phase. The frame size is also n'. Let P_1 be the probability that no other tag is mapped to the slot that x is mapped to. The subscript '1' indicates that a single hash function is used.

$$P_1 = \left(1 - \frac{1}{n'}\right)^{n'-1} \approx e^{-\frac{n'-1}{n'}} \approx e^{-1} \approx 36.8\%, \tag{3.1}$$

where e is the natural constant. When this happens, tag x will be assigned to the slot. It will not participate in the remaining phases. Hence, the expected number of phases that tag x participates is

$$1 \times P_1 + 2 \times (1 - P_1)P_1 + 3 \times (1 - P_1)^2 P_1 + \dots$$

$$= \sum_{i=0}^{\infty} P_1(1 - P_1)^i + \sum_{i=1}^{\infty} P_1(1 - P_1)^i + \sum_{i=2}^{\infty} P_1(1 - P_1)^i + \dots$$

$$= 1 + (1 - P_1) + (1 - P_1)^2 + \dots$$

$$= \frac{1}{P_1} \approx e,$$

where we have used the fact that $\sum_{i=0}^{\infty} P_1(1 - P_1)^i = 1$. Recall that in any phase the number of slots in the time frame is equal to the number of tags considered for slot assignment. In other words, each time x participates in a phase, its presence contributes a slot in the frame. Overall, the expected number of slots in all phases that can be attributed to x's participation is e. The total number of tags in the system is n. Therefore, the number of slots in the frames of all phases is expected to be $n \times e$. The expected frame time in all phases is $e \times n \times t_{inf}$.

There is a one-to-one correspondence between bits in an indicator vector and slots in a frame. Hence, the total number of bits in all indicator vectors is also $n \times e$. The expected time for transmitting all indicator vectors is $e \times n \times t_{id}/96$. Due to the large denominator of 96, it is smaller than the total frame time, $e \times n \times t_{inf}$. The rest of the information collection request excluding the indicator vector is very small and can be ignored. Hence, the expected execution time of SIC is $e \times n \times t_{inf} + e \times n \times t_{id}/96$. The first item is about 2.72 times of the lower bound $n \times t_{inf}$.

We go back to (3.1). Out of the n' slots in a frame, the number of useful slots is $n'P_1 \approx 36.8\% \times n'$. Hence, in each phase of the SIC protocol, only about 36.8% of the time slots are useful and 63.2% of the slots are wasted. This gives us significant room for further improvement.

3.4 Multi-hash Information Collection Protocol

We present a Multi-hash Information Collection protocol (MIC) to solve the hash collision problem of SIC.

3.4.1 Protocol Overview

MIC is similar to SIC except that it assigns tags to slots using k hash functions in order to alleviate the problem of wasted slots. More specifically, it hashes each tag to k slots in the frame. As long as any one of these slots has no other tag, the reader is able to assign the tag to the slot.

The execution of the MIC protocol also consists of multiple phases. In each phase, the RFID reader broadcasts an information collection request that is followed by a slotted time frame, in which some tags transmit their information.

In the following, we first explain how to assign tags to slots by using k hash functions. We then introduce a mechanism (called hash-selection vector) to inform the tags about the assignment.

3.4.2 Assigning Tags to Slots Using Multiple Hash Functions

The k hash functions are denoted as $H[i]$, $1 \le i \le k$, which takes a tag ID and a random number r as input and produces a pseudo-random number $H[i](ID, r)$ as output.

Consider an arbitrary phase in the execution of MIC. Let n' be the number of tags for slot assignment in this phase. The frame size is also set to be n'. If it is the first phase, $n' = n$.

The slot assignment consists of k rounds, each involving one hash function. In the first round, we apply $H[1]$ to map the tags to the slots. A tag is assigned to a slot if it is the only one that is mapped to the slot. After assignment, we remove the tag from being further considered in the remaining rounds that involve other hash functions. We also mark the slot as being *occupied*. The slots that are not marked at the end of this round are said to be *unoccupied*.

In the second round, we apply $H[2]$ to map the remaining tags to the n' slots. If a tag is mapped to an unoccupied slot and it is the only one that is mapped to the slot, it is assigned to the slot and removed from further consideration. The slot is marked as occupied.

The above process is repeated for other hash functions one round after another in order to assign as many tags as possible to the unoccupied slots. An illustrative example is given in Fig. 3.1 (a)-(b). After all k hash functions are used, the reader has a subset of tags that are assigned to slots, and it also know which hash function each of these tags should use. The problem of communicating this information to the tags will be addressed shortly.

Please be aware of the difference between the term "map" and the term "assign". A tag may be mapped to multiple slots based on the k hash functions, but it can only be assigned to one slot. Also be aware of the difference between "phase" and "round". Each execution of MIC consists of phases. Each phase is a request-and-response exchange between the reader and the tags. Before that exchange, the reader has to assign tags to slots and the assignment process consists of multiple rounds when more than one hash function is used.

3.4.3 Protocol Description

In the first phase of the MIC execution, before sending out an information collection request, the RFID reader determines which tags are assigned to which slots (see the previous subsection). It then constructs an n-element *hash-selection vector*. Each element in the vector corresponds to a slot in the frame at the same index location. If no tag is assigned to a slot, the reader sets the corresponding element in the hash-selection vector to zero. If a tag is assigned to a slot using the jth hash function, the reader sets the corresponding element to be j. The size of an element is $\lceil \log_2(k+1) \rceil$ bits.

The request broadcast by the reader consists of the information type to be reported, the frame size, a random number r, and the hash-selection vector, where r is used by the hash functions and it is different in each phase. The hash-selection vector is divided into segments of 96 bits (equivalent to the length of a tag ID), and each segment is transmitted in a time slot of size t_{id}.

The tags will receive the hash-selection vector along with other information in the request. Using the same k hash functions, each tag knows which k slots in the frame and which k elements in the hash-selection vector it is mapped to. If a tag is assigned to a slot, it must be one of those k slots. If a tag is assigned to a slot by the reader using the jth hash function, the corresponding element in the hash-selection vector must have a value of j because this is exactly how the hash-selection vector is constructed. The inverse is also true. *If a tag finds that (1) it is mapped to a slot s using the jth function and (2) the corresponding element in the hash-selection vector is also j, then it can conclude that it must have been assigned to slot s by the reader. If multiple hash functions satisfy the above conditions, the tag only uses the one that has the smallest value of j.* See Sect. 3.4.6 for correctness proof. An illustrative example is given in Fig. 3.1.

Hence, in order to determine whether it is assigned to a slot, a tag only needs to examine the elements in the hash-selection vector that it is mapped to by the k hash functions. Let E_j, $1 \le j \le k$, be the element that the tag is mapped to by the jth hash function. The tag examines the elements in order from E_1 to E_k. If it finds the value of an element E_j is equal to j, the tag knows that it must be assigned to a slot by the

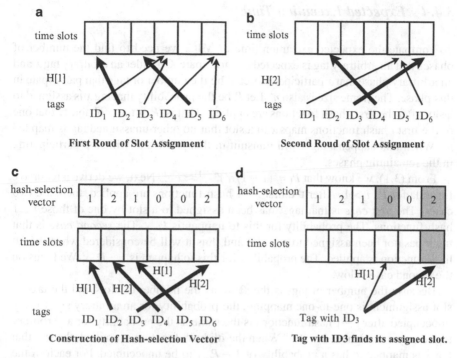

a

time slots

H[1]

tags   ID$_1$ ID$_2$ ID$_3$ ID$_4$ ID$_5$ ID$_6$

First Roud of Slot Assignment

b

time slots

H[2]

tags   ID$_1$ ID$_2$ ID$_3$ ID$_4$ ID$_5$ ID$_6$

Second Roud of Slot Assignment

c

hash-selection vector

| 1 | 2 | 1 | 0 | 0 | 2 |

time slots

H[1]

H[1]   H[2]   H[2]

tags   ID$_1$ ID$_2$ ID$_3$ ID$_4$ ID$_5$ ID$_6$

Construction of Hash-selection Vector

d

hash-selection vector

| 1 | 2 | 1 | 0 | 0 | 2 |

time slots

H[1]   H[2]

Tag with ID3

Tag with ID3 finds its assigned slot.

Fig. 3.1 Arrows represent the mapping from tags to slots based on hash functions. Among them, thick arrows represent the assignment of tags to slots. In this example, $k = 2$. (**a**) Two tags, ID$_1$ and ID$_5$, are assigned to slots in the first round when $H[1]$ is applied. (**b**) Two more tags, ID$_3$ and ID$_4$, are assigned in the second round when $H[2]$ is applied. (**c**) The reader constructs a hash-selection vector based on the slot assignment. (**d**) After receiving the vector, the tag with ID$_3$ examines the elements in the hash-selection vector that it is mapped to. The element mapped by $H[2]$ has a value of 2. The tag knows that it must be assigned to the corresponding slot.

reader using the jth hash function. In this case, it will stop examining the remaining elements and wait until the assigned slot arrives. It will transmit during that slot and then stop participating further in the protocol execution.

Note that a tag does not have to receive the whole hash-selection vector. It knows the indices of the elements it looks for. The tag can be in a stand-by mode to conserve energy at times other than it receives its segments of the hash-selection vector or transmits its information.

After the first phase completes, the RFID reader moves to the second phase, which is identical except that the reader removes the tags for which it has assigned slots. It only considers the tags that have not got a chance to transmit. The frame size in this phase is reduced accordingly.

The above process repeats phase after phase until all tags report their information. Alternatively, the reader may stop when the number of remaining tags is fewer than a small threshold, and it invokes the PIC protocol to collect information from these tags.

Clearly, SIC is a special case of MIC when $k = 1$.

3.4.4   Expected Execution Time

To compute the expected execution time of MIC, we need to find the number of phases that an arbitrary tag is expected to participate. Consider an arbitrary tag x and an arbitrary phase that x participates. Let n' be the number of tags that participate in this phase. The frame size is also n'. Let P_i be the probability that tag x is assigned to a slot after the first l hash functions are applied. That is, P_i is the probability that one of the first i hash functions maps x to a slot that no other unassigned tag is mapped to. When this happens, tag x will transmit in this phase and will stop participating in the remaining phases.

From (3.1), we know that $P_1 = (1 - 1/n')^{n'-1} \approx e^{-1}$. Next, we derive a recursive formula for P_i, $i > 1$. After the first $i - 1$ hash functions are applied, there are two cases. The *first case* is that tag x has been assigned to a slot by one of those $i - 1$ hash functions. The probability for this to happen is P_{i-1}. The *second case* is that tag x has not been assigned to any slot and thus it will be considered when the ith hash function is applied. The probability for this to happen is $1 - P_{i-1}$. We focus on the second case below.

Because the number of tags is the same as the number of slots and the tag-to-slot assignment is one-to-one mapping, the probability for an arbitrary slot to stay unoccupied after $i - 1$ hash functions is the same as the probability for an arbitrary tag to stay unassigned, $1 - P_{i-1}$. When the ith hash function is applied, the slot that tag x is mapped to has a probability of $1 - P_{i-1}$ to be unoccupied. For each of the other $n' - 1$ tags, it has a probability of $1 - P_{i-1}$ to be unassigned and, if so, it has a probability of $\frac{1}{n'}$ to be mapped to the same slot as x does. Hence, the probability p for tag x to be the only one that is mapped to an unoccupied slot is

$$p = (1 - P_{i-1})\left(1 - (1 - P_{i-1})\frac{1}{n'}\right)^{n'-1} \approx (1 - P_{i-1})e^{-(1-P_{i-1})} \qquad (3.2)$$

Recall that we are considering the second case here. Combining both cases discussed previously, we have

$$P_i = P_{i-1} + (1 - P_{i-1}) \times p = P_{i-1} + (1 - P_{i-1})^2 e^{-(1-P_{i-1})}, \qquad (3.3)$$

where the first item on the right side is the probability for a tag to be assigned to a slot by one of the first $i - 1$ hash functions and the second item is the probability for the tag to be assigned to a slot by the ith hash function. The probability for tag x to be assigned to a slot after all k functions are applied is P_k, and this is the case for any phase that x participates.

Based on the recursive formula in (3.3), we compute the numerical values of P_i in Table 3.1, which match perfectly with the simulation results in Sect. 3.6. If seven hash functions are used, i.e., $k = 7$, the probability for an unassigned tag to be assigned to a slot in any phase is $P_7 \approx 86.1\%$. The probability for an arbitrary slot to be useful is also 86.1%. Only 13.9% of the slots in each frame is wasted.

Table 3.1 Numerical values of P_i, which can be interpreted as the probability for any slot to be useful.

P_1	P_2	P_3	P_4	P_5	P_6	P_7
36.8%	58.0%	69.6%	76.4%	80.8%	83.9%	86.1%

The expected number of phases that tag x participates is

$$1 \times P_k + 2 \times (1 - P_k)P_k + 3 \times (1 - P_k)^2 P_k + \ldots = \frac{1}{P_k}.$$

For each phase that x participates, the reader allocates a slot in the frame. Hence, the expected number of slots that can be attributed to x's participation in the protocol is $1/P_k$. There are n tags. The total number of slots in all phases is expected to be n/P_k. The total expected time in the frames of all phases is $n \times t_{inf}/P_k$. When $k = 7$, it becomes $1.16 \times n \times t_{inf}$, just 16% more than the lower bound. Because 32 elements of the hash-selection vector can fit in a segment of 96 bits, the expected time for all indicator vectors is $n \times t_{id}/(32P_k)$. Hence, the expected execution time of MIC is about $n \times t_{inf}/P_k + n \times t_{id}/(32P_k)$.

The ratio of $n \times t_{id}/(32P_k)$ to the lower bound $n \times t_{inf}$ is largest when the information reported by each tag is one bit. In this case, t_{id} is about 12 times of t_{inf}, according to the parameters in [9]. Hence, when $k = 7$, $n \times t_{id}/(32P_k)$ is up to 45% of the lower bound. Consequently, the expected execution time of MIC is up to 1.61 times the lower bound.

3.4.5 Hash Functions

There are many efficient hash functions in the literature. We describe a simple implementation that helps to keep the complexity of a tag's circuit low. The tags do not have to fully implement the k hash functions, $H[i](ID, r)$. When $k = 1$, the expected number of phases a tag will participate is just $1/P_1 \approx 2.7$, which means that a tag only needs to produce 2.7 hash values on average. Similarly, when $k = 3$, a tag needs $1/P_3 \approx 1.4$ hash values on average. When $k = 7$, a tag needs $1/P_7 \approx 1.2$ hash values on average. For $n = 50,000$, each hash value is 16 bits long. We may derive these hash values from a ring of pre-stored random bits as follows: We use an offline random number generator with the ID of a tag as seed to generate a string of random bits. We take k segments of the string. Each segment contains a certain number of bits, forming a ring by logically connecting the last bit with the first bit. These rings are pre-stored in the tags before they are deployed. To find the value of $H[i](ID, r)$, a tag takes a certain number of bits from the ith ring. More specifically, it takes a number of bits from the ring clockwise beginning from the rth bit. An alternative approach is to begin from the rth bit and take one bit after every r bits until a sufficient number of bits are taken. The final hash value is the number represented by these bits modulo the frame size.

The larger the size of each ring, the better the pseudo randomness in the hash output. Because the protocols we present only requires a tag to produce a very small number of hash values from each ring, a ring size of 100 bits should be more than sufficient. In this case, when $k = 3$, each tag needs to store 300 bits to implement the hash functions. When $k = 7$, each tag needs 700 bits.

The RFID reader knows the IDs of the tags, and it picks the random number r. Hence, it can predict the hash values of all tags.

3.4.6 Correctness

In Sect. 3.4.2, the RFID reader assigns a tag to a slot only when no other tag is mapped to the same slot. After a tag is assigned to a slot, the reader removes it from further consideration. Hence, from the reader's point of view, each tag is uniquely assigned to a slot. According to the protocol description in Sect. 3.4.3, each tag will only transmit once. What we want to make sure is that the tag will transmit in the assigned slot. Moreover, there should not be collision in that slot.

To determine in which slot it transmits, a tag first uses the k hash functions to map itself to k slots in the frame. The rule states that: *If the tag finds that (1) it is mapped to a slot s using the jth function and (2) the corresponding element in the hash-selection vector is also j, then it can conclude that it must have been assigned to slot s by the reader. If multiple hash functions satisfy the above conditions, the tag only uses the one that has the smallest value of j.* When a tag x is mapped to a slot using the jth hash function, if the tag finds that the corresponding element in the hash-selection vector is also j, it means that the reader has assigned a tag to the slot based on the jth function, and moreover the reader will do so only when a single tag is mapped to the slot using the jth function. Tag x can thus conclude that this single tag must be itself. Because it is the only tag that will transmit in this slot, there will not be a collision.

3.5 Frame-Optimized Information Collection Protocol

In this section, we present an optimization of MIC. The optimized protocol is called the *frame-optimized information collection protocol* (FIC).

3.5.1 Motivation

In the design of SIC or MIC, we set the frame size f to be the number n' of tags from which the RFID reader has not yet received their information. However, it appears arbitrary to set f to be n' even though good performance has been achieved.

A natural question is whether we can achieve better performance by choosing a frame size different from n'. Indeed, we have observed in the simulations that when f is chosen smaller than n', the protocol execution time may be noticeably reduced.

We present a variant of MIC, called the frame-optimized information collection protocol (FIC). It is identical to MIC except that before each phase, the RFID reader makes a query for the optimal frame size (which is pre-computed and stored in a table) and broadcasts the frame size as part of the information collection request. In the following, we will explain how to compute the optimal frame size and discuss the performance of FIC.

3.5.2 Computing Optimal Frame Size

The execution of FIC also consists of multiple phases. Consider an arbitrary phase. Let n' be the number of tags from which the reader has not yet received information. We derive the functional relationship between the frame size f and the percentage of time slots that will be useful. Based on this functional relationship, we can find the value of f that maximizes the percentage of slots that are useful.

Consider an arbitrary participating tag x. In FIC, we use P'_i to denote the probability that tag x is assigned to a slot after the first i hash functions are applied. That is, P'_i is the probability that one of the first i hash functions maps x to a slot that no other unassigned tag is mapped to. When this happens, tag x will transmit during this phase of FIC and will stop participating in the remaining phases.

Let $P'_0 = 0$. We derive a recursive formula for P'_i, $i \geq 1$. After the first $i - 1$ hash functions are applied, there are two cases. In the first case, tag x is assigned to a slot by one of those $i - 1$ hash functions. The probability for this to happen is P'_{i-1}. In the second case, tag x has not been assigned to any slot and thus it will be considered when the ith hash function is applied. The probability for this to happen is $1 - P'_{i-1}$.

Under the condition of the second case, let p' be the probability for x to be assigned a slot by the ith hash function. We have

$$P'_i = P'_{i-1} + (1 - P'_{i-1}) \times p'. \tag{3.4}$$

Next, we derive p'. For tag x to be assigned a slot, it has to be mapped by the ith function to an unoccupied slot, and no other unassigned tag is mapped to the same slot. After the previous hash functions are applied, each tag has a probability of P'_{i-1} to be assigned a slot. Since there are n' tags and f slots, it translates into a probability of $P'_{i-1} \times n'/f$ for any slot to be occupied. Hence, the probability for x to be mapped to an unoccupied slot is $(1 - P'_{i-1} \times n'/f)$.

Consider any other tag. The probability that it is not yet assigned to a slot is $(1 - P'_{i-1})$. Under that condition, the probability that it is mapped to the same slot as x does is $1/f$. The probability that the tag is either already assigned to a slot or mapped to a different slot than x is $(1 - (1 - P'_{i-1})/f)$. Since there are $n' - 1$ tags other than x, the probability for all of them to be either assigned to slots already or mapped to different slots is $(1 - (1 - P'_{i-1})/f)^{n'-1}$.

Fig. 3.2 Value of $P'_k \times n'/f$ with respect to the frame size f when $k = 7$ and $n' = 50,000$.

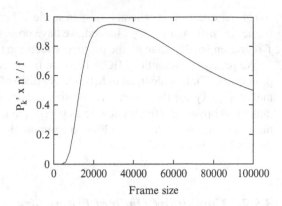

Based on the above analysis, the probability p' for tag x to be the only one that is mapped to an unoccupied slot is

$$p' = \left(1 - P'_{i-1}\frac{n'}{f}\right)\left(1 - (1 - P'_{i-1})\frac{1}{f}\right)^{n'-1}$$

$$\approx \left(1 - P'_{i-1}\frac{n'}{f}\right)e^{-(1-P'_{i-1})\frac{n'}{f}}. \tag{3.5}$$

Applying it to (3.4), we have

$$P'_i = P'_{i-1} + (1 - P'_{i-1})\left(1 - P'_{i-1}\frac{n'}{f}\right)e^{-(1-P'_{i-1})\frac{n'}{f}}, \tag{3.6}$$

where the first item on the right side is the probability for a tag to be assigned to a slot by one of the first $i - 1$ hash functions and the second item is the probability for the tag to be assigned to a slot by the ith hash function. Since there are n' tags and f slots, the probability for an arbitrary slot to be occupied by a tag after i hash functions is $P'_i \times n'/f$. The probability for an arbitrary slot to be useful (i.e., occupied by a tag) after all k functions are applied is $P'_k \times n'/f$.

We want to find the optimal value of f that maximizes $P'_k \times n'/f$. As we increase the value of f, on one hand, it has a negative impact on the value of $P'_k \times n'/f$ because f is in the denominator. On the other hand, it also has a positive impact because P'_k is an increasing function of f; the probability for a tag to be assigned to a slot is certainly higher when there are more slots available for assignment. Our numerical computation shows that when f is small, the positive impact dominates and thus the value of $P'_k \times n'/f$ increases. When f is large, the negative impact dominates and $P'_k \times n'/f$ decreases. For an example, we plot the functional relationship between $P'_k \times n'/f$ and f in Fig. 3.2, where $n' = 50,000$ and $k = 7$. It shows that $P'_k \times n'/f$ is an increasing function at first and then a decreasing function. The optimal value of f that maximizes $P'_k \times n'/f$ is 29,617.

Algorithm 1 Bisection search for the optimal value of f.

$f_1 = 1, f_2 = F$
while $f_2 - f_1 > 1$ **do**
  $\bar{f} = \lfloor f_1 + f_2 \rfloor / 2$
  if $\frac{P'_k \times n'}{\bar{f}} < \frac{P'_k \times n'}{\bar{f}+1}$ **then**
    $f_1 = \bar{f}$
  else
    $f_2 = \bar{f}$
  end if
end while
if $\frac{P'_k \times n'}{f_1} < \frac{P'_k \times n'}{f_2}$ **then**
  return f_2
else
  return f_1
end if

Fig. 3.3 Optimal value of f with respect to n' when $k = 7$.

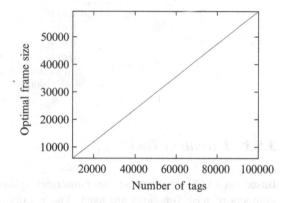

We can use numerical methods to compute the optimal value of f. One method is exhaustive search: It tries different values of f, starting from 1 and increasing f by one each time until further increment in f causes $P'_k \times n'/f$ to decrease. A more efficient method is bisection search: See Algorithm 1. We set f_1 to a value (such as 1) that is smaller than the optimal value of f. Set f_2 to a value (denoted as F) that is larger than the optimal value of f. The empirical computation shows that n' is always larger than the optimal value of f. Hence, we may let $F = n'$. We repeat the following operation: Let $\bar{f} = \lfloor f_1 + f_2 \rfloor / 2$. If $P'_k \times n'/\bar{f} < P'_k \times n'/(\bar{f}+1)$, set f_1 to be $\bar{f}$; otherwise, set f_2 to be $\bar{f}$. The above iterative operation stops when $f_2 - f_1 \leq 1$.

The optimal value of f with respect to n' is computed before hand and stored in a table on the RFID reader or a server. Before the reader broadcasts an information collection request, it looks up in the table for the optimal value of f under the current value of n'. It then includes f in the broadcast. When $k = 7$, the optimal value of f is plotted in Fig. 3.3.

Table 3.2 Maximum value of $P_k \times n'/f$, which is the probability for an arbitrary slot to be useful. The frame size that achieves the maximum value of $P_k \times n'/f$ is also shown.

k	$P'_k \times n'/f$ (or C_k)	f
1	36.8%	49999
2	59.3%	41045
3	73.5%	36385
4	82.7%	33601
5	88.8%	31784
6	92.7%	30526
7	95.2%	29617

Fig. 3.4 Value of $P'_k \times n'/f$ with respect to n' when $k = 7$ and the optimal frame size f is used.

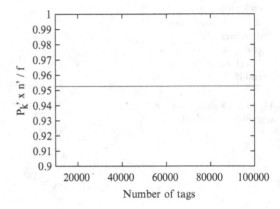

3.5.3 Execution Time

Based on (3.6), we compute the numerical values of $P'_k \times n'/f$ when different numbers of hash functions are used. The results are shown in Table 3.2, where $n' = 50,000$. We use the optimal frame sizes. They are shown in the third row. Recall that $P'_k \times n'/f$ is the probability for an arbitrary slot to be useful. It can also be interpreted as the percentage of all time slots that are useful. For example, when $k = 7$, 95.2% of all slots are useful, which compares favorably to 86.1% that is achieved by MIC (see P_7 in Table 3.1).

Through extensively numerical computations, we find that when k is fixed, $P'_k \times n'/f$ is largely insensitive to n'. For example, as shown in Fig. 3.4 where $k = 7$, the value of $P'_k \times n'/f$ can be treated as a constant. Let's denote the constant as C_k. For example, $C_7 = 95.2\%$.

As FIC performs its multiple phases, the percentage of time slots that are useful is C_k in each phase. The total number of useful slots in all phases is equal to the number n of tags in the system. Hence, the total number of slots in all phases must be n/C_k. Therefore, the execution time of FIC is $n \times t_{inf}/C_k + n \times t_{id}/(32C_k)$, where $n \times t_{inf}/C_k$ is the total time of all slots and $n \times t_{id}/(32C_k)$ is the total time for the reader to transmit the hash-selection vector.

When $k = 7$, the ratio of $n \times t_{inf}/C_k$ to the lower bound $n \times t_{inf}$ is 1.05. The ratio of $n \times t_{id}/(32C_k)$ to the lower bound is largest when the information reported by each tag is one bit. In this case, t_{id} is about 12 times of t_{inf}, according to the parameters in [9]. Hence, when $k = 7$, $n \times t_{id}/(32C_k)$ is up to 39% of the lower bound. Consequently, the execution time of FIC is up to 1.44 times the lower bound.

3.5.4 Channel Error

We now consider the impact of channel error. If a segment in the hash-selection vector is corrupted, the tags that extract information from that segment may transmit in wrong slots, causing collision. It may also happen that the information transmitted by a tag in the correct slot is corrupted by noise in the channel. If the channel error is mild and the application can tolerate a certain level of error, the protocol may not need additional error control mechanisms. For example, suppose the reader collects the battery status of the tags to see if any battery needs to be replaced. The reader may periodically collect such information. Over a period of time, it will receive a certain number of readings from each tag. It decides whether a tag needs to replace battery based on the majority votes. In this way, occasional information corruption due to channel error does not cause a misjudgement of the battery status. In another example, consider a large chilled food storage facility and the application is to monitor the temperature at each section of the storage by using sensor-augmented RFID tags that are attached to the food items. Each section has many tags, which provide a large amount of redundancy in the information reported to the RFID reader. If temperature readings from some tags are corrupted, the reader can still retrieve correct temperature data by removing outliers from all the readings it receives from a particular section of the storage.

If the application requires that the information received from every tag is correct, we need to add checksum such as a CRC code to each transmission for error detection. Each segment of 96 bits in the hash-selection vector carries 16-bit checksum, and it uses the remaining 80 bits to carry 26 elements of 3 bits each. Each information report from a tag also carries 16-bit checksum. Consider the following two cases: (1) When a tag finds that one of its k segments in the hash-selection vector is corrupted, it ignores the segment and only uses other segments to decide whether it is assigned to a slot. If none of the other segments suggests that it is assigned to a slot, it makes the conservative decision that it will not participate in the remaining phases even through it does not transmit in this phase, because the reader might have assigned it to a slot using the corrupted segment. (2) Beside the case that a tag may not transmit at all, even when a tag transmits in the correct slot, that slot may be corrupted due to channel error, which can be detected by the reader through the mismatching CRC code.

The reader handles the above cases in the same way: It only assigns one slot to each tag during the execution of FIC (MIC or SIC). If it does not receive a tag's information in the assigned slot or the information is corrupted, it will not assign another slot because otherwise we would run into the issue of acknowledging the tags whether their information is received correctly — this can get complicated, considering that the acknowledgement itself may also be corrupted. After FIC (MIC or SIC) completes, the reader wakes up all tags and performs the polling protocol (PIC) on the set of tags from which the information has not been received correctly. We expect the set to be relatively small in a practical environment where the channel noise is not too large to hinder the effectiveness of the RFID system. In PIC, each transmission also carries CRC. Due to channel error, the reader may poll for the information of a tag more than once until the information is correctly received.

3.6 Simulation Results

We now compare the execution time of protocols through simulations with and without channel errors.

3.6.1 Simulation Setting

The simulation setting is based on the Philips I-Code specification [9]. Any two consecutive transmissions (from the reader to tags or vice versa) are separated by a waiting time of 302 μs. According to the specification, the transmission rate from a tag to the reader is different than the transmission rate from the reader to a tag. The rate from a tag to the reader is 53 Kb/sec; it takes 18.88 μs for a tag to transmit one bit. The value of t_{inf} is calculated as the sum of a waiting time and the time for transmitting the information, which is 18.88 μs multiplied by the length of the information. For example, if the sensor information is one bit, t_{inf} is 321 μs if a CRC code is not added, and it is 623 μs if a 16-bit CRC is added. If the sensor information is 16 bits, t_{inf} is 604 μs without CRC, and it is 906 μs with CRC.

The transmission rate from the reader to tags is 26.5 Kb/sec; it takes 37.76 μs for the reader to transmit one bit. Each tag ID contains 96 bits, which include a 16-bit CRC code according to the Gen2 standard. Recall that t_{id} is the time it takes the reader to transmit an ID to a tag. It is 3927 μs (including a waiting time before the transmission). The time for the reader to transmit a segment of the indicator vector or a segment of the hash-selection vector is the same. However, if a tag transmits a 96-bit ID to the reader, it only takes 2114 μs due to a different transmission rate.

In each simulation run, we set the number n of tags in the RFID system. We then execute seven protocols: PIC, SIC, MIC with $k = 3$, MIC with $k = 7$, FIC with $k = 3$, FIC with $k = 7$, and EDFSA [4]. EDFSA is one of the best ID-collection protocols. We modify it for information collection. The modification is simple: When a tag transmits its ID to the reader, it piggybacks its sensor information. We measure and compare the execution times of the five protocols. Each data point in the figures is the average outcome of 100 simulation runs under the same setting.

Table 3.3 Execution time comparison (in seconds) when the sensor information is 1 bit long.

	n = 10,000	30,000	50,000	70,000	90,000
EDFSA	90.5	261.9	587.7	771.9	1021.0
PIC	42.5	127.4	212.4	297.3	382.3
SIC	9.9	30.0	49.2	69.2	89.2
MIC, k = 3	5.8	17.5	28.9	40.6	52.2
MIC, k = 7	5.2	15.5	25.8	36.2	46.4
FIC, k = 3	5.5	16.2	27.1	37.9	48.6
FIC, k = 7	4.7	14.0	23.2	32.4	41.7
lower bound	3.2	9.6	16.0	22.5	28.9

In the following, we first present the simulation results when CRC codes are not used for error control, and then we present the results when CRC codes are used (see Sect. 3.5.4).

3.6.2 Execution Time Comparison

We first study the performance of the protocols without considering channel error. That is, the RFID reader can always correctly receive the information when a tag transmits in a slot without collision.

Table 3.3 compares the execution times of the seven protocols under different values of n when the sensor information is one bit. This corresponds to the application of monitoring the battery status of the RFID tags: '1' means the battery is ok; '0' means the battery needs to be replaced. We examine the fourth column in the table for $n = 50,000$ (imagine that a large military base stores 50,000 pieces of weapons and ammunition packets, each attached with a tag). The execution time of EDFSA is 530.1 seconds, which is about thirty-three times of the lower bound, 16.0 seconds. PIC reduces the execution time by 60% to 212.4 seconds because it eliminates collisions that exist in EDFSA, which is ALOHA-based. SIC further reduces the time to 49.7 seconds, about one fourth of the time needed by PIC. MIC reduces the execution time to 29.0 seconds when three hash functions are used or 25.8 when seven hash functions are used. The best protocol, FIC, is able to further reduce the execution time to 23.2 when seven hash functions are used. Similar conclusions can be drawn from other columns: FIC works the best, MIC is the next, SIC follows, then PIC, and finally EDFSA.

Table 3.4 and Table 3.5 present the execution times when the sensor information is 16 bits long and 32 bits long, respectively. Again, similar conclusions can be drawn. For example, when the information is 16 bits long and $n = 50,000$, the execution time of FIC with $k = 7$ is 43% of the time needed by SIC, 17% of the time needed by PIC, and just 6.6% of the time needed by EDFSA. When the information is 32 bits long and $n = 50,000$, the execution time of FIC with $k = 7$ is 41% of the time needed by SIC, 22% of the time needed by PIC, and 8.2% of the time needed by EDFSA.

Table 3.4 Execution time comparison (in seconds) when the sensor information is 16 bits long.

	n = 10,000	30,000	50,000	70,000	90,000
EDFSA	92.1	293.1	575.7	815.7	1150.0
PIC	45.3	135.9	226.6	317.2	407.8
SIC	17.8	53.2	88.0	124.3	158.5
MIC, k = 3	9.9	29.7	49.2	69.3	88.7
MIC, k = 7	8.5	25.4	42.2	59.2	76.1
FIC, k = 3	9.2	27.6	45.9	64.1	82.5
FIC, k = 7	7.6	22.8	37.9	53.0	68.3
lower bound	6.0	18.1	30.2	42.3	54.4

Table 3.5 Execution time comparison (in seconds) when the sensor information is 32 bits long.

	n = 10,000	30,000	50,000	70,000	90,000
EDFSA	106.8	303.7	645.0	951.5	1252.2
PIC	48.3	145.0	241.7	338.3	435.0
SIC	25.5	78.0	129.9	183.1	232.3
MIC, k = 3	14.2	42.7	70.9	99.5	127.8
MIC, k = 7	11.9	35.9	59.8	83.8	107.6
FIC, k = 3	13.3	39.5	65.9	92.5	119.5
FIC, k = 7	10.8	32.2	53.7	75.1	96.6
lower bound	9.1	27.2	45.3	63.4	81.6

Table 3.6 Execution time comparison (in seconds) when the channel error rate is 1%.

	n = 10,000	30,000	50,000	70,000	90,000
EDFSA	80.5	253.4	435.0	657.2	921.7
PIC	46.4	139.3	232.1	325.0	417.8
SIC	19.3	57.9	96.6	136.4	173.9
MIC, k = 3	11.3	34.0	56.4	79.2	101.5
MIC, k = 7	10.0	29.6	49.4	69.1	88.9
FIC, k = 3	10.6	31.8	52.9	74.0	95.4
FIC, k = 7	9.0	27.0	44.8	62.7	80.6

The execution time of FIC with $k = 3$ is only slightly worse than that of FIC with $k = 7$. Because each tag has to store k hash outputs, if one wants to reduce the storage overhead, k may be chosen as 3.

3.6.3 Execution Time Comparison under Channel Error

The method for handling channel error is described in Sect. 3.5.4. Suppose the sensor information is 1 bit long. Tables 3.6–3.8 present the execution time comparison when the channel error rate is 1%, 5% and 10%, respectively. The channel error rate c is defined as the percentage of slots that is corrupted. In the

Table 3.7 Execution time comparison (in seconds) when the channel error rate is 5%.

	n = 10,000	30,000	50,000	70,000	90,000
EDFSA	84.6	248.9	498.8	651.1	926.0
PIC	50.4	151.2	252.1	352.9	453.7
SIC	23.3	69.7	117.1	162.6	209.7
MIC, k = 3	15.4	45.7	76.8	107.0	137.6
MIC, k = 7	14.0	41.6	69.4	97.1	124.8
FIC, k = 3	14.6	43.6	72.8	102.1	130.9
FIC, k = 7	13.0	38.8	64.8	90.6	116.5

Table 3.8 Execution time comparison (in seconds) when the channel error rate is 10%.

	n = 10,000	30,000	50,000	70,000	90,000
EDFSA	89.8	252.6	489.3	738.7	1070.2
PIC	56.1	168.5	280.8	393.2	505.5
SIC	29.4	87.4	144.8	203.4	261.3
MIC, k = 3	21.1	63.1	105.1	147.4	189.3
MIC, k = 7	19.6	58.9	98.2	137.3	176.7
FIC, k = 3	20.3	61.0	101.8	142.3	183.0
FIC, k = 7	18.7	56.1	93.5	130.9	168.3

simulations, each slot has a probability of c to be corrupted. With the presence of different levels of channel error, we continue to observe that FIC outperforms MIC, MIC performs much better than SIC, SIC is better than PIC, which is in turn better than EDFSA.

For example, in Table 3.6 where the channel error rate is 1%, when $n = 50,000$, the execution time of FIC with $k = 7$ is 44.8seconds, the time of MIC with $k = 7$ is 49.4seconds, the time of SIC is 96.9seconds, the time of PIC is 232.1seconds, and the time of EDFSA is 460.0seconds. In Table 3.7 where the channel error rate is 5%, when $n = 50,000$, the execution time of FIC with $k = 7$ is 64.8seconds, the time of MIC with $k = 7$ is 69.4seconds, the time of SIC is 116.8seconds, the time of PIC is 252.1seconds, and the time of EDFSA is 480.1seconds. In Table 3.8 where the channel error rate is 10%, when $n = 50,000$, the execution time of FIC with $k = 7$ is 93.5seconds, the time of MIC with $k = 7$ is 98.1seconds, the time of SIC is 145.6seconds, the time of PIC is 280.8 seconds, and the time of EDFSA is 505.8seconds.

3.6.4 Values of P_i and $P'_k \times n'/f$

To verify the analytical results in Sect. 3.4.4, we measure the values of P_i by simulations. As shown in Table 3.9, the values of P_i measured from simulations match well with the numerically-computed values based on the recursive formula (3.3), which confirms the correctness of the analysis.

Table 3.9 values of P_i

	P_1	P_2	P_3	P_4	P_5	P_6	P_7
by simulation	37.0%	58.2%	69.9%	76.6%	80.8%	83.8%	86.2%
by analysis	36.8%	58.0%	69.6%	76.4%	80.8%	83.9%	86.1%

Table 3.10 values of $P_k' \times n'/f$

k	1	2	3	4	5	6	7
by simulation	36.8%	59.3%	73.5%	82.7%	88.8%	92.7%	95.2%
by analysis	36.9%	59.1%	73.2%	82.8%	89.0%	92.4%	95.1%

Similarly, to verify the analytical results in Sect. 3.5.2 and 3.5.3, we measure the values of $P_k' \times n'/f$ by simulations. Shown in Table 3.10, the values of $P_k' \times n'/f$ measured from simulations match the computed values from the analysis.

3.7 Summary

This chapter investigates the problem of efficiently collecting sensor information from all tags to a reader in a large RFID system. We present three protocols. The first one, called the polling-based information collection protocol (PIC), serves as a baseline for comparison. The second protocol, called the single-hash information collection protocol (SIC), improves time and energy efficiencies by totally eliminating the transmission of tag IDs. It uses a hash function to assign tags to the slots of a frame, during which the tags can transmit their data successfully. However, due to hash collisions, many slots have to be wasted. A wide gap still exists between the execution time of SIC and a lower bound that we establish. We use multiple hash functions to solve the hash collision problem, which leads to the third protocol, called the multi-hash information collection protocol (MIC). Its execution time is about half of the execution time of the SIC, up to seven times smaller than the execution time of PIC, and up to nineteen times smaller than the execution time of a representative ID-collection protocol [4] that is enhanced to collect sensor information. An optimized version of MIC, called the frame-optimized information collection protocol (FIC), works even better. Its execution time is within 1.44 times the lower bound that we have established.

References

1. Information Technology – Radio Frequency Identification for Item Management Air Interface – Part 6: Parameters for Air Interface Communications at 860-960 MHz. Final Draft International Standard ISO 18000-6 (2003)
2. Bhandari, N., Sahoo, A., Iyer, S.: Intelligent Query Tree (IQT) Protocol to Improve RFID Tag Read Efficiency. Proc. of IEEE ICIT (2006)

3. Cha, J.R., Kim, J.H.: Dynamic Framed Slotted ALOHA Algorithms using Fast Tag Estimation Method for RFID Systems. Proc. of IEEE CCNC (2006)
4. Lee, S., Joo, S., Lee, C.: An Enhanced Dynamic Framed Slotted ALOHA Algorithm for RFID Tag Identification. Proc. of IEEE MOBIQUITOUS (2005)
5. Myung, J., Lee, W.: Adaptive Splitting Protocols for RFID Tag Collision Arbitration. Proc. of ACM MOBIHOC (2006)
6. Ni, L.M., Liu, Y., Lau, Y.C., Patil, A.: LANDMARC: Indoor Location Sensing using Active RFID. ACM Wireless Networks (WINET) 10(6) (2004)
7. Roberts, L.G.: ALOHA Packet System with and without Slots and Capture. ACM SIGCOMM Computer Communication Review 5(2), 28–42 (1975)
8. Sarangan, V., Devarapalli, M.R., Radhakrishnan, S.: A Framework for Fast RFID Tag Reading in Static and Mobile Environments. The International Journal of Computer and Telecommunications Networking 52(5) (2008)
9. Semiconductors, P.: I-CODE Smart Label RFID Tags. http://www.nxp.com/documents/data_sheet/SL092030.pdf(2004)
10. Sheng, B., Li, Q., Mao, W.: Efficient Continuous Scanning in RFID Systems. Proc. of IEEE INFOCOM (2010)
11. Vogt, H.: Efficient Object Identification with Passive RFID Tags. Proc. of IEEE PerCom (2002)
12. Zhen, B., Kobayashi, M., Shimizu, M.: Framed ALOHA for Multiple RFID Objects Identification. IEICE Transactions on Communications (2005)
13. Zhou, F., Chen, C., Jin, D., Huang, C., Min, H.: Evaluating and Optimizing Power Consumption of Anti-collision Protocols for Applications in RFID Systems. Proc. of ISLPED (2004)

Chapter 4
Tag-ordering Polling Protocols in RFID Systems

4.1 System Model

We consider a large RFID system using active tags. Each tag carries a unique ID and one or more sensors. It also has the capability of performing certain computations as well as communicating with the RFID reader wirelessly. The reader and the tags transmit with sufficient power such that they can communicate over a long distance. We assume that the RFID reader knows the IDs of all tags in the system by executing an ID-collection protocol, and it has enough power supply.

Let N be the set of tags in the system and $n = |N|$. Let M be a subset of tags, $m = |M|$, and $M \subseteq N$. The objective is to design efficient polling protocols that collect information from tags in M. A polling protocol may be scheduled to execute periodically. M may change over time so that different subsets of tags are queried. We have two performance objectives. The primary performance objective is to achieve *energy efficiency*. We want to minimize the average amount of energy that a tag spends during one execution of a polling protocol. The energy expenditure by a tag has two components: (1) energy for transmitting its information (e.g., 32 bits) to the reader, and (2) energy for receiving the polling request and other information from the reader. The former is a small, fixed amount of energy that must be spent. The latter is dependent on the protocol design as we will see shortly. It is a variable amount of energy that should be minimized. Simple protocol designs will result in all tags in the system, including those not in M, to be continuously active and unnecessarily receive a large amount of data from the reader for an extended period of time. How to minimize such energy cost is the focus of this chapter.

The secondary performance objective is to reduce *protocol execution time*. RFID systems use low-rate communication channels. For example, in the Philips I-Code system, the rate from a reader to a tag is about 27 Kbps and the rate from a tag to a reader is about 53 Kbps. Low rates, coupled with a large number of tags, often cause long execution times for RFID protocols. To apply such protocols in a busy warehouse environment, it is desirable to reduce protocol execution time as much as possible.

Y. Qiao et al., *RFID as an Infrastructure*, SpringerBriefs in Computer Science, DOI 10.1007/978-1-4614-5230-0_4, © The Author(s) 2013

Communication between the reader and tags is time-slotted. The reader's signal synchronizes the clocks of tags. Let t_{id} be the length of a time slot during which the reader is able to broadcast a tag ID, and t_{inf} be the length of a time slot during which a tag is able to transmit its information.

4.2   Basic Polling Protocol (BP)

In a standard, straightforward way of designing a polling protocol, we simply let the RFID reader broadcast the tag IDs in M one by one. After it transmits an ID, it waits for a time slot of t_{inf} during which the corresponding tag transmits its information. Each tag continuously listens to the wireless channel. Whenever it receives an ID from the reader, the tag compares the received ID with its own ID. If they match, the tag will transmit its information and then go to sleep until the next scheduled execution of the protocol.

In the above protocol, each tag in M will have to receive $m/2$ IDs on average from the reader before it transmits. Each tag not in M will have to receive all m IDs. The amount of energy spent by a tag in receiving such data grows linearly with respect to m. It takes a constant amount of energy for a tag to receive an ID and another constant amount of energy for it to transmit its information. The energy cost of the whole system is thus $O(nm)$. The protocol execution time is $m(t_{id} + t_{inf})$.

We use a numerical example to explain the energy cost. Consider a military base that has a large warehouse storing 50,000 weapons, ammunition magazines, and other equipment, which are tagged with RFID sensors. Among them, there are 1,000 sensitive devices, from which a RFID reader needs to access information in order to make sure that they are in good conditions or simply to confirm their presence (against unauthorized removal). Let e_r be the amount of energy a tag spends in receiving an ID and e_s be the amount of energy a tag spends in transmitting its information. The total energy consumed by all tags for transmitting is $1,000e_s$, and the total energy consumed by all tags for receiving is about $50,000,000e_r$. Even though e_r may be smaller than e_s, the total amount of energy spent by tags in receiving can be much greater than the amount spent in transmitting.

4.3   Coded Polling Protocol (CP)

We show that a coded polling protocol (CP) [4] is able to reduce the amount of data each tag has to receive by half. The protocol assumes that each tag ID carries an *identification number* and a CRC (*cyclic redundancy code*) for error detection. This requirement is satisfied by the EPCglobal Gen-2 standard, where each 96-bit tag ID contains a CRC checksum. The CRC is computed based on the identification number and a generator. When a tag receives an ID from a wireless channel, it computes a CRC based on the received identification number and then compares the result with the received CRC. If they are the same, we say the ID contains a *valid* CRC.

CRC has the following property: If x and y are two tag IDs with valid CRCs, then $x \oplus y$ also has a valid CRC. The same property does not hold for $x \oplus \hat{y}$, where $\hat{y}$ contains the same bits in y but in the reverse order. For example, if $y = 10110$, then $\hat{y} = 01101$. We call $\hat{y}$ the *reversal* of y.

In the coded polling protocol, the RFID reader first arranges the IDs in M in pairs. Each pair consists of two IDs that are arbitrarily selected from M. Consider an arbitrary pair, x and y, which are called each other's *paring ID*. We define the *polling code* of the pair as $c = x \oplus \hat{y}$.

Instead of sending out the IDs in M one after another, the reader broadcasts the polling code of each pair one after another. After each broadcast of a polling code $c = x \oplus \hat{y}$, the reader waits for two time slots, during which tag x and tag y will transmit. More specifically, when an arbitrary tag z receives the polling code c, it first computes $z \oplus c$, and checks whether the CRC in the reversal of $z \oplus c$ is valid. If it is, the tag will transmit its information. Otherwise, the tag computes $\hat{z} \oplus c$, and checks whether the CRC in $\hat{z} \oplus c$ is valid. Again, if it is valid, the tag will transmit. Otherwise, the tag will not transmit. We show that only tag x and tag y will transmit.

First, consider the case of $z = x$. The tag first computes $z \oplus c = x \oplus x \oplus \hat{y} = \hat{y}$. The reversal of $\hat{y}$ is y. The CRC in any tag ID (including y) is valid. Hence, tag x will transmit. Moreover, it now knows its pairing ID, y. If x is greater than y, the tag will transmit in the first slot after receiving the polling code; otherwise, it will transmit in the second slot.

Second, we consider the case of $z = y$. The tag first computes $y \oplus c = y \oplus x \oplus \hat{y}$. Its reversal is likely to have an invalid CRC; the chance for an arbitrary number to contain a valid CRC is very small. Then, the tag computes $\hat{z} \oplus c = \hat{y} \oplus x \oplus \hat{y} = x$, which contains a valid CRC. Consequently, y will transmit. Since it now knows its pairing ID, x, it also knows in which slot it should transmit.

Finally, consider the case of $z \neq x$ and $z \neq y$. The tag computes the reversal of $z \oplus c = z \oplus x \oplus \hat{y}$ and then computes $\hat{z} \oplus c = \hat{z} \oplus x \oplus \hat{y}$. Both of them are likely to have invalid CRCs.

A minor problem is that $y \oplus c$ in the second case and $z \oplus c$ or $\hat{z} \oplus c$ in the third case still have a small probability to contain a valid CRC. However, the reader can easily prevent this from happening. It knows all tag IDs. It can precompute all polling codes and check whether a valid CRC happens in the above cases by chance when it is not supposed to. If this is true for a pair of tags, x and y, the reader must break up the pair, and use them to form new pairs with other IDs in M. Such an approach is effective because the probability for this to happen is exceedingly small when CRC is sufficiently long.

Because each polling code represents two tag IDs, the number of polling codes in CP is $m/2$. Hence, when comparing with the basic polling protocol, CP reduces the number of broadcasts made by the reader by half, and it also reduces the amount of data that each tag has to receive by half. This not only saves energy for tags, but also reduces the protocol execution time to $mt_{id}/2 + mt_{inf}$.

4.4 Tag-Ordering Polling Protocol (TOP)

Although CP is more efficient, the expected amount of energy that each tag spends in receiving remains $O(m)$. In this section, we present a new tag-ordering polling protocol that reduces such energy cost to $O(1)$.

4.4.1 Motivation

In the basic polling protocol, a RFID reader broadcasts m IDs in time slots of length t_{id}. All tags must continuously monitor the wireless channel in order to know whether their own IDs are in the broadcast. In CP, the reader broadcasts $m/2$ polling codes also in time slots of length t_{id}. Again, all tags must continuously monitor the wireless channel. They have to keep receiving and processing the polling codes. Each tag in the basic protocol has to receive up to m IDs. Even though CP is more efficient, a tag still has to receive up to $m/2$ codes.

We want to remove the necessity for any tag to keep monitoring the wireless channel. Ideally, a tag should stay in an energy-conserving standby mode for most of time, and only wake up at the right time slot to receive information about itself, such as whether it is polled and, if so, when it should transmit. To further reduce the amount of data that tags have to receive, we let the reader broadcast a so-called *reporting-order* vector V, instead of IDs in M. Each ID in M is mapped to a bit in V through a hash function; the bit is set as one to encode the ID in the vector. A tag only needs to check a specific bit in V at a location determined by the hash of its ID. This bit is called the *representative bit* of the tag. If its value is one, the tag is polled by the reader for reporting, i.e., the tag belongs to M; if its value is zero, the tag is not polled. The vector V also carries information about the order in which the polled tags will report their data. Each bit whose value is one in V represents a polled tag. If a tag finds that there are i ones in V preceding its representative bit, it knows that it should be the $(i+1)$th tag in M to report its information. With such an ordering, it becomes possible for tags in M to report at different times and avoid collision.

However, this basic idea has two problems. First, there should be at least m bits in V to encode m IDs in M. The energy cost of receiving V remains $O(m)$. How can a tag find out the number of ones in V preceding its representative bit without having to receive the whole vector? Second, hash collision causes two issues. If a tag not in M is hashed to the same bit in V as a tag in M does, it will find its representative bit to be one, causing false positive. If two tags in M are mapped to the same bit in V, they will transmit at the same time, causing report collision. In the rest of this section, we design a new tag-ordering polling protocol (TOP) to solve these problems.

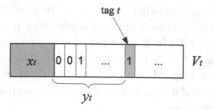

Fig. 4.1 V_t is the representative segment of tag t, x_t is the total number of ones in all previous segments, and y_t is the number of ones in V_t that precede tag t's representative bit. I_t is the position of t in the reporting order. $I_t = x_t + y_t$.

4.4.2 Protocol Description

TOP consists of three phases: *ordering phase*, *polling phase*, and *reporting phase*. In the ordering phase, the reader broadcasts the vector V so that each tag knows whether it is polled and where it is located in the reporting order. The polling phase resolves the issues of false positive and report collision. Finally, in the reporting phase, tags in M report their information in the order defined by V without collision.

4.4.2.1 Ordering phase

The RFID reader does not broadcast any IDs or indices. It only broadcasts the reporting-order vector, V. If V cannot fit in one time slot of length t_{id}, the reader breaks the vector into segments and broadcasts each segment in a time slot of t_{id}. In addition, the reader also broadcasts the vector size v.

Knowing the vector size, a tag t is able to hash its ID and find out the location of its representative bit in V. Because the segment size is fixed, t also knows which segment its representative bit belongs to. This segment, denoted as V_t, is called the *representative segment* of tag t. A tag will stay in the standby mode and be active only when receiving its representative segment.

If a tag finds that its representative bit is zero, it knows for sure that it is not a member in M. If a tag finds that its representative bit is one, it may be a member in M or a non-member that is mapped to a bit which a member in M is also mapped to. The latter case causes *false positive*. Because the reader knows all IDs in the system, it can pre-compute the set F of non-member tags that cause false positive.

When the reader broadcasts any segment of V, it includes in the same time slot *the total number of ones in the previous segments*. For an arbitrary tag t, let I_t be the number of ones in V preceding the representative bit of t. When tag t receives V_t, it can computes I_t as the sum of (a) the number of ones in the previous segments and (b) the number of ones in V_t before its representative bit. see Fig. 4.1 for illustration. As we will see later, the value of I_t specifies when tag t will transmit during the reporting phase.

If two tags in M are mapped to the same bit in V, they will have the same I_t value and thus transmit at the same time during the reporting phase, causing collision. Because the reader has all IDs in M, it knows exactly which tags will be mapped to the same bit. This makes it easy to resolve collision. The reader simply removes all but one tag that are mapped to a bit, and puts them in a set C. These tags, together with tags in F, will not participate in the reporting phase. They are handled separately in the polling phase.

4.4.2.2 Polling Phase

In this phase, the reader issues two types of polling requests. For each tag in C, it sends a *positive polling request*. For each tag in F, it sends a *negative polling request*. To distinguish these two types, the reader must transmit a one-bit flag together with a tag ID in each request, specifying whether the polling is positive or negative and which tag is polled.

Tags that find their representative bits to be ones in the previous phase must continuously listen to the channel during the polling phase. After sending a positive request, the reader waits for a time slot to receive information. The tag that finds its ID in the request will transmit its information in this slot. This tag, which belongs to C, will not participate in the reporting phase. After sending a negative request, the reader does not wait before sending out the next request. The tag that finds its ID in a negative request knows that it must belong to F and hence should not further participate in the protocol execution.

The total number of polling requests is $|F| + |C|$. By choosing an appropriate size for the reporting-order vector, we show later that we can make sure $|F| + |C| = O(1)$. Note that only tags in M and F have to listen to the channel in this phase. Tags in $N - M - F$, which may contain the majority of tags in the system, have already known that they do not belong to M and thus do not need to participate in the protocol execution.

4.4.2.3 Reporting phase

A tag participates in the reporting phase only if it satisfies the following two conditions: (1) it finds that its representative bit is one in the ordering phase, and (2) it does not find its ID in the requests of the polling phase.

The reporting phase consists of $m - |C|$ time slots. In each time slot, one tag in $M - C$ transmits its information. Recall that each tag in M learns its index in the reporting order during the ordering phase. The tag will transmit in the reporting phase at the time slot of the same index.

4.4.2.4 Timing

Before executing the protocol, the RFID reader uses its broadcasting signal to synchronize the clocks of the tags. The reader computes the vector V and breaks it into segments. Suppose each time slot of length t_{id} can carry 96 bits. We may set the segment size to be 80 bits and use the remaining 16 bits to carry the total number of ones in the previous segments.[1] The reader is able to compute the execution time T_1 of the ordering phase, which is the number of segments multiplied by t_{id}.

Since the reader knows all IDs in the system, it can precompute the set F of tags that cause false positive and the set C of tags that should not participate in the reporting phase in order to avoid collision. Based on F and C, the reader can compute the execution time T_2 of the polling phase, which is $|F| \times t_{id} + |C| \times (t_{id} + t_{inf})$.

Suppose all tags wake up at each scheduled execution of the protocol. The reader computes and broadcasts the values of T_1 and T_2 right before the ordering phase, so that the tags know when each phase of the protocol will begin. They will remain in the standby mode unless they have to receive their representative segments, participate in the polling phase, or transmit their information in the reporting phase.

If the system requires on-demand polling of tag information instead of periodic execution, there are two possible solutions to wake the tags up in the first place. The first one is "pseudo-on-demand" polling, where tags still wake up periodically, but the reader only issues the polling request when needed. The second approach is to attach a wake-up circuit to each tag, and use the two-stage wake-up scheme proposed in [5] to activate the tags. In this approach, tags responde almost immediately to the polling event. However, the wake-up circuit requires the reader to be close enough so that the radio power is strong enough to trigger the wake-up event. As a result, we may have to deploy extra readers to cover all the tags.

4.5 Performance Analysis of TOP

We show the energy cost and execution time of TOP through analysis in this section. Also, the energy-time trade-off is analyzed and the time-constrained energy minimization problem is discussed.

---

[1]Using 16 bits to carry the number of ones in previous segments will limit the value of m to $(0, 65,535]$. To get rid of this limitation, we can use $\lceil \log_2 m \rceil$ bits instead and broadcast the value of $\lceil \log_2 m \rceil$ to tags at the beginning of protocol. However, for the sake of simplicity, we use 16 bits in this chapter to help demonstrate the main idea.

4.5.1 Energy Cost

We show how to configure TOP such that the energy cost per tag is $O(1)$. The energy cost of a tag has four components: (1) receiving v, T_1 and T_2, (2) receiving a segment of V in the ordering phase, (3) listening to the channel during the polling phase, and (4) transmitting information in a slot at the reporting phase (or at the polling phase if the tag is in C). The first two components incur small, constant energy expenditure to every tag in the system. The fourth component also incurs small, constant energy cost, but only to the tags in M. The third component incurs energy cost only to tags in F and M. In the worse case, a tag has to listen to all $|C| + |F|$ polling requests from the reader. Suppose it takes one unit of energy to receive a polling request. The total energy cost of a tag, denoted as Ω, is

$$\Omega \leq |C| + |F| + O(1). \tag{4.1}$$

We treat $|C|$ and $|F|$ as random variables and derive their expected values. Recall that v be the number of bits in the reporting-order vector V. Let b_i be the value of the ith bit in V, $0 \leq i < v$. For each tag in M, the reader maps it to a random bit in V and sets the bit to one. After encoding all m tags in V, the probability for b_i to be one is

$$Prob\{b_i = 1\} = 1 - \left(1 - \frac{1}{v}\right)^m \approx 1 - e^{-m/v}. \tag{4.2}$$

The bits, $b_0, b_2, ..., b_{v-1}$, are independent of each other. Thus, the expected number of ones in V is $\sum_{i=1}^{v} Prob\{b_i = 1\}$. The value of $|C|$ is equal to m subtracted by the number of ones in V. Hence, we have

$$E(|C|) = m - \sum_{i=1}^{v} Prob\{b_i = 1\} \approx m - v\left(1 - e^{-m/v}\right). \tag{4.3}$$

A tag not in M will cause false positive when its representative bit is one. The probability for this to happen is $Prob\{b_i = 1\}$. Hence,

$$E(|F|) = (n - m)Prob\{b_i = 1\} \approx (n - m)\left(1 - e^{-m/v}\right). \tag{4.4}$$

Both $E(|C|)$ and $E(|F|)$ are monotonically decreasing functions of v. We show that $E(|C|) = O(1)$ if v is sufficiently large. Let $v = m^2/2$. From Taylor expansion, we know that

$$1 - e^{-m/v} = \frac{m}{v} - \frac{1}{2!}\left(\frac{m}{v}\right)^2 + \frac{1}{3!}\left(\frac{m}{v}\right)^3 - \frac{1}{4!}\left(\frac{m}{v}\right)^4 \cdots$$

$$\geq \frac{m}{v} - \frac{1}{2!}\left(\frac{m}{v}\right)^2.$$

Applying it to (4.3), we have

$$E(|C|) = m - v\left(1 - e^{-m/v}\right) \le \frac{1}{2!}\frac{m^2}{v} = 1. \tag{4.5}$$

Next we show that $E(|F|) = O(1)$ if v is sufficiently large. If $n = m$, $E(|F|) = 0$. Now assume $n > m$. Let $v = -m/\ln[1 - 1/(n - m)]$. Applying it to (4.4), we have

$$E(|F|) = (n - m)\left(1 - e^{-m/v}\right) = 1. \tag{4.6}$$

Therefore, if we choose $v = \max\{m^2/2, -m/\ln[1 - 1/(n - m)]\}$, we have

$$E(\Omega) \le E(|C|) + E(|F|) + O(1) \le 1 + 1 + O(1) = O(1).$$

We conclude that TOP can be configured such that the expected energy cost per tag is $O(1)$. As we will see shortly, the protocol execution time increases when v becomes too large. To strike a balance between energy cost and protocol execution time, we may choose a value of v much smaller than $\max\{m^2/2, -m/\ln[1 - 1/(n - m)]\}$. Later we will use simulations to study the performance of TOP under practical values of v. For example, when $v = 24m$, the amount of data that a tag receives in TOP is more than an order of magnitude smaller than what a tag has to receive in CP.

We characterize the energy cost in the polling phase by counting the amount of data (in Kilobits) that a tag has to receive. Numerical results are shown in the first plot of Fig. 4.2, where $n = 50,000$ and $m = 5,000, 10,000$, or $25,000$, corresponding to three curves in the plot. Clearly, as v increases, the energy cost decreases.

4.5.2   Execution Time

The protocol execution time also consists of four components. To begin with, it takes the reader a small, constant time to broadcast v, T_1 and T_2. The time for the ordering phase is vt_{id}/l, where l is the segment size. The time for the polling phase is $|F| \times t_{id} + |C| \times (t_{id} + t_{inf})$. The time for the reporting phase is $|M - C| \times t_{inf}$. Hence, the total execution time is

$$T = \left(\frac{v}{l} + |F| + |C|\right) t_{id} + m \times t_{inf} + O(1). \tag{4.7}$$

From (4.3) and (4.4), the expected protocol execution time is

$$E(T) = \left[\frac{v}{l} + (n - m)\left(1 - e^{-m/v}\right) + m - v(1 - e^{-m/v})\right] t_{id} + m \cdot t_{inf} + O(1)$$

$$\approx \left[\frac{v}{l} + \frac{(n - m)m}{v}\right] t_{id} + m \cdot t_{inf} + O(1). \tag{4.8}$$

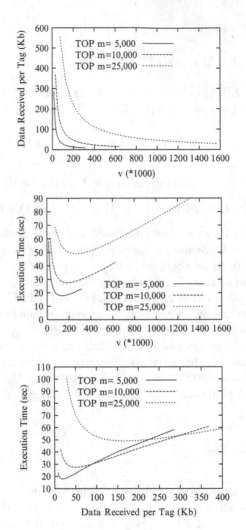

Fig. 4.2 Energy cost and execution time of TOP. **First plot**: Energy cost per tag with respect to v. **Second plot**: Protocol execution time with respect to v. **Third plot**: Energy-time tradeoff controlled by v.

The second plot of Fig. 4.2 presents the protocol execution time (excluding the constant $O(1)$) when $n = 50,000$, $m = 5,000, 10,000$, or $25,000$, $t_{id} = 3297\mu s$, and $t_{inf} = 906\mu s$; see Sect. 4.9 for how they are determined. Interestingly, as v increases, the execution time first decreases and then increases. We can find the optimal value of v that minimizes the execution time from $\frac{\delta E(T)}{\delta v} = 0$.

Combining the results in the first and second plots, we can figure out the *tradeoff relation* between energy cost and protocol execution time, which is presented in the third plot. As v becomes large, the energy cost decreases at the expense of increased execution time.

4.5.3   Choosing v for Time-constrained Energy Minimization

Recall the performance objectives of TOP are energy efficiency and time efficiency. However, as shown in Fig. 4.2, we may not be able to achieve the best performance in both metrics using one configuration. Below we study how to configure TOP for time-constrained energy minimization.

Consider a warehouse with a large number of RFID-tagged goods. Suppose the system administer wants to maximize the tags' battery lifetime, but there is a requirement on the execution time of a polling operation because excessively long execution time increases the chance of interfering with other scheduled tasks. From the previous analysis, we know that the protocol execution time is treated as a random variable. Let T be the execution time of TOP, B be a pre-defined time bound, and α be a probability value, $0 < \alpha < 1$. The time constraint can be specified in a probabilistic way,

$$Prob\{T \leq B\} \geq \alpha. \tag{4.9}$$

Our performance objective is to find the optimal value of v that minimizes the energy cost, subject to the above constraint.

As shown in the first plot of Fig. 4.2, the energy cost decreases as the size of the reporting-order vector, v, increases. Hence, the goal becomes finding the largest v that satisfies (4.9). In the following, we derive $Prob\{T \leq B\}$ as a function of v. Based on this function, we will be able to compute the optimal value of v.

Let d be the total number of ones in V after encoding tags in M, $0 < d \leq m$. The probability that x bits are ones, expressed as $Prob\{d = x\}$, can be calculated by the *balls and bins algorithm*, which will be given in the next subsection. For now we denote the function for computing $Prob\{d = x\}$ as $p_d(m, v, x)$.

After encoding tags in M, the reader removes colliding tags to C. The value of $|C|$ is equal to m subtracted by the number of ones in V. Hence,

$$Prob\{|C| = c\} = Prob\{d = m - c\} = p_d(m, v, m - c). \tag{4.10}$$

When a tag not in M is mapped to a bit that is one, false positive happens. The reader puts all false positive tags to F. When there are x bits that are ones in V, the conditional false positive probability is x/v. Thus,

$$Prob\{\text{false positive} \mid d = x\} = \frac{x}{v}.$$

Obviously, when $d = x$, the total number of false positive tags follows a binomial distribution $Bino(n - m, x/v)$.

$$Prob\{|F| = f \mid d = x\} = \binom{n - m}{f} \left(\frac{x}{v}\right)^f \left(1 - \frac{x}{v}\right)^{n - m - f}.$$

Let S be the union of C and F, so $|S| = |C| + |F|$. The probability distribution of $|S|$ is

$$Prob\{|S| = s\} = \sum_{c=0}^{s} Prob\{|F| = s - c \mid |C| = c\} \cdot Prob\{|C| = c\}$$

$$= \sum_{x=1}^{m} Prob\{|F| = s - m + x \mid d = x\} \cdot Prob\{d = x\}$$

$$= \sum_{x=1}^{m} \binom{n-m}{s-m+x} \left(\frac{x}{v}\right)^{s-m+x} \left(1 - \frac{x}{v}\right)^{n-s-x} p_d(m, v, x). \quad (4.11)$$

Adopt (4.7) and ignore $O(1)$, a small constant time for the reader to broadcast v, T_1 and T_2, which is negligibly small when comparing with other components on the right side of (4.7). We have

$$Prob\{T \leq B\} = \sum_{s=0}^{s_{max}} Prob\{|S| = s\}, \quad (4.12)$$

where $s_{max} = (B - mt_{inf})/t_{id} - v/l$. We denote the right side of (4.12) as $P_t(v, B)$, which is the probability for the protocol execution time to be bounded by B under a certain value of v. It is computable as a function of v and B after (4.11) is applied and parameters m and n are given.

We want to find the largest value of v that satisfies the inequality, $P_t(v, B) \geq \alpha$. Our numerical computation shows that, given a fixed value of B, $P_t(v, B)$ is not a monotonic function with respect to v. Hence, we cannot directly apply the bisection search method to find the largest v that satisfies $P_t(v, B) \geq \alpha$. We may use the False Position algorithm [3] to find the optimal value of v. The computation overhead is reasonable. For $n = 10,000$, $m = 1,000$, $B = 4$ seconds, and $\alpha = 99\%$, it takes an Apple macbook (2.4GHz CPU and 4GB memory) 3 seconds to find the optimal $v = 60,160$. And for $n = 10,000$, $m = 1,000$, $B = 3$ seconds, and $\alpha = 99\%$, it takes the same computer 16 seconds to find that no v can satisfy the requirement, because $B = 3$ seconds is smaller than the minimum execution time that TOP can achieve.

As a related problem, if v and α are given, we can also use $P_t(v, B)$ to compute the time bound that TOP can achieve. More specifically, given a value of v, we are able to find the smallest B that satisfies $P_t(v, B) \geq \alpha$ through bisection search: Recall that $P_t(v, B)$ is the formula for $Prob\{T \leq B\}$, the probability for the protocol execution time to be bounded by B. Clearly, it is an increasing function of B with $P_t(v, 0) = 0$ and $P_t(v, +\infty) = 1$. We choose a small value B_1 (e.g., 0) such that $P_t(v, B_1) < \alpha$ and a large value B_2 such that $P_t(v, B_2) \geq \alpha$. Let $B_3 = \lceil (B_1 + B_2)/2 \rceil$. If $P_t(v, B_3) < \alpha$, assign B_3 to B_1; otherwise, assign B_3 to B_3. Hence, the search range $[B_1, B_2]$ is cut by half. Repeat the above process until $B_1 = B_2$, which gives the smallest bound B that satisfies $P_t(v, B) \geq \alpha$. Let $n = 10,000$ and $m = 1,000$. Figure 4.3 shows the smallest bound B with respect to v when $\alpha = 90\%, 95\%$ and 99%, which correspond to the three curves in the figure.

Fig. 4.3 Bound B that
satisfies $Prob\{T \leq B\} \geq \alpha$
with respect to v. Parameters:
$n = 10,000, m = 1,000$.

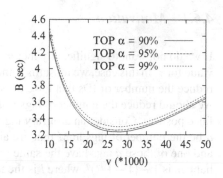

4.5.4 Computing $p_{d(m,\,v,\,x)}$ — the Balls and Bins Algorithm

Problem: Suppose we throw m balls into v empty bins. Each ball is thrown to a random bin, and each bin can hold unlimited number of balls. We want to find the probability that after m balls are thrown, x bins are not empty, denoted as $p_d(m,v,x)$.

Solution: There are many solutions to this problem. We now provide a recursive one. Assume after we throw m balls, there are x non-empty bins, $1 \leq x \leq m$. When $x > 1$, there are two possibilities of where the mth ball goes: (1) If the mth ball is placed to a previously empty bin, there should be $x-1$ non-empty bins after $m-1$ balls were thrown, and the possibility for this to happen is $(v-x+1)/v$; (2) Otherwise if the mth ball goes to a previously non-empty bin, there must be x non-empty bins after $m-1$ balls were thrown, and the possibility of this option is x/v. Thus,

$$p_d(m,v,x) = \begin{cases} 1; x = m = 1. \\ \frac{x}{v}p_d(m-1,v,x) + \frac{v-x+1}{v}p_d(m-1,v,x-1); 1 \leq x \leq m \text{ and } x \leq v. \\ 0; \text{ all other cases.} \end{cases}$$

$p_d(m,v,x)$ can be calculated from simple dynamic programing.

4.6 Enhanced Tag-Ordering Polling Protocol (ETOP)

In this section, in present a protocol (ETOP) that will further improve the performance of TOP. We first show that there is ;still space for improvement in TOP and then give the protocol description.

4.6.1 Motivation

If we do not want to significantly increase execution time, we cannot choose a large value for v. In this case, we must find other means to lower energy cost. The key is to reduce the number of IDs that have to be transmitted in the polling phase. Namely, we should reduce the number of tags in F and C. Let's first focus our discussion on false positive. Consider an arbitrary tag $t \notin M$. Its representative segment is V_t. Let q be the number of tags in M that are also mapped to V_t. False positive occurs if t and one of those q tags have the same representative bit. The probability for this to happen is $1 - (1 - 1/l)^q$, where l is the number of bits in V_t.

To further reduce the false-positive probability, we can implement each segment of V as a Bloom filter [1, 2]. The reader uses multiple hash functions to map each tag to $k(> 1)$ *representative bits* in V, instead of just one in TOP. More specifically, for each member $t' \in M$, the reader first maps it to a representative segment $V_{t'}$ through a hash function whose range is $[0, v/l)$. Then the reader further maps t' to k representative bits in $V_{t'}$ and set them to ones.

After all members in M are encoded in the segments of V, the reader broadcasts the segments in the ordering phase. A tag t only listens for its representative segment V_t and then checks its representative bits. If any representative bit is zero, the tag can not be in M. If all representative bits are ones, the tag may be a member in M or a false positive. In the case of false positive, even though the tag does not belong to M, every one of its representative bits is set because it is also a representative bit of a member tag in M. The probability for this to happen is $(1 - (1 - 1/l)^{kq})^k$, where q is the number of tags in M whose representative segments are also V_t. For example, if $l = 80$, $k = 3$, and $q = 2$, the false-positive probability is just 3.8×10^{-4}, much lower than $1 - (1 - 1/l)^q = 2.5 \times 10^{-2}$ in TOP under the same parameters.

Bloom filters can reduce the false-positive probability. But it is more difficult to use them to carry the reporting order, based on which the tags will take turn to transmit during the reporting phase. In TOP, we use the number of ones that precede the representative bit of a tag to determine the tag's position in the reporting order. Bloom filters use multiple representative bits to encode each member. The representative bits of different members may overlap in an arbitrary way. Hence, we cannot simply use all bits whose values are ones to represent tags in M because there is no one-to-one mapping between them.

In the following, we design an *enhanced tag-ordering polling protocol* (ETOP) to solve the above problem. ETOP uses *partitioned Bloom filters*, which not only reduce false positive and encode the reporting order, but also reduce $|C|$ as well as overall execution time of the protocol.

Fig. 4.4 V_t is the representative segment of tag t. V_t is evenly divided into k partitions, each having $\lfloor l/k \rfloor$ bits. Tag t has one representative bit in every partition.

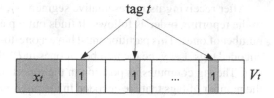

4.6.2   Protocol Description

The main difference between ETOP and TOP is that ETOP implements each segment of V as a partitioned Bloom filter instead of a simple bit array. When we describe the protocol of ETOP, we focuses on the difference while omitting the details that it shares in common with TOP.

In a partitioned Bloom filter, the l bits of a segment are evenly divided into k partitions. Each partition has $\lfloor l/k \rfloor$ bits. see Fig. 4.4 for illustration. For every member tag t in M, the reader applies a hash function on its ID to obtain a number of hash bits. The reader uses $\lceil \log_2 v \rceil$ hash bits to map t to a representative segment V_t, and then uses $k \lceil \log_2 l/k \rceil$ hash bits to further map t to one representative bit in every partition of the segment. Like a classical Bloom filter, the partitioned Bloom filter sets k representative bits for each encoded member; unlike a classical Bloom filter, a partitioned Bloom filter spreads the k representative bits in k different partitions.

After receiving its representative segment, a tag checks the k representative bits to determine if it is a member in M. False positive cases are handled by the reader in the polling phase as usual.

How does a tag t know its position in the reporting order? First we consider the reporting order among tags that are encoded in the same segment V_t. Since every tag has exactly one representative bit in each partition of V_t, we may be able to use one of the partitions to carry the order information. In other words, if there is a partition P^* whose number of ones is equal to the number of tags encoded in V_t, we know that there must be a one-to-one mapping between these tags and the '1' bits in P^*. We can use the order of '1' bits in P^* as the reporting order of the corresponding tags. We will explain later how the reader makes sure that such a partition exists. When the reader sends out V_t, in the same time slot it also sends the total number x_t of tags that are encoded in all previous segments of V. The position of tag t in the reporting order can be computed from x_t and the information in P^*, which we will further explain shortly.

How to make sure that any segment of V always has a partition whose number of ones is equal to the number of tags encoded in the segment? The reader has to do some extra work. After encoding all tags in M, the reader examines the partitions one by one for each segment. If there is not such a partition, the reader removes an encoded tag and places it in the set C, which will be explicitly polled in the polling phase. The reader keeps removing tags until it finds a partition that satisfies the above requirement. Note that the requirement is always satisfied when the number of tags encoded in a segment is one.

After receiving its representative segment V_t, a tag $t \in M$ computes its position in the reporting order as follows: It finds out a partition P^* in V_t that has the largest number of ones. This partition must have a one-to-one mapping between '1' bits and encoded tags. Let y_t be the number of ones in P^* that precedes the representative bit of t. The tag computes its position in the reporting order as $y_t + x_t$. Recall that x_t is the number of tags that are encoded in the previous segments. It is received together with V_t in the same time slot.

The polling phase and the reporting phase of ETOP are identical to their counterparts in TOP.

We show the energy cost and execution time of ETOP through analysis in this section. Also, the energy-time trade-off is analyzed and the time-constrained energy minimization problem is discussed.

4.7 Performance Analysis of ETOP

4.7.1 Energy Cost

We show that ETOP can be configured such that the energy cost per tag is $O(1)$. ETOP has the same upper bound formula for per-tag energy cost as TOP does, which is shown in (4.1), but it has different values of $|C|$ and $|F|$. In the following, we derive $|C|$ and $|F|$ for ETOP. Let m_i be the number of tags in M that are encoded in the ith segment, $0 \le i < v/l$. Each tag in M has a probability of l/v to be mapped to the ith segment. Hence, m_i follows a binomial distribution $Bino(m, l/v)$.

$$Prob\{m_i = x\} = \binom{m}{x} \left(\frac{l}{v}\right)^x \left(1 - \frac{l}{v}\right)^{m-x}. \tag{4.13}$$

Let C_i be a subset of C, containing the tags that are removed from the ith segment. We know the following facts: (1) When $m_i = 0$, $|C_i| = 0$. (2) When $m_i = 1$, $|C_i| = 0$. (3) When $m_i \ge 1$, $|C_i| \le m_i - 1$. Hence, we must have

$$E(|C_i|) < (m_i - 1) \cdot \left(1 - Prob\{m_i = 0\} - Prob\{m_i = 1\}\right)$$

$$= (m_i - 1) \cdot \left(1 - \left(1 - \frac{l}{v}\right)^m - \frac{ml}{v}\left(1 - \frac{l}{v}\right)^{m-1}\right).$$

Since $(1 - l/v)^m > 1 - ml/v$, we have

$$E(|C_i|) < \frac{m_i(m-1)^2 l^2}{v^2} < \frac{m_i m^2 l^2}{v^2}.$$

$|C|$ is the sum of all $|C_i|$s, $0 \le i < v/l$. We know $\sum_{i=1}^{v/l} m_i = m$. So,

$$E(|C|) = \sum_{i=1}^{v/l} E(|C_i|) < m\frac{m^2 l^2}{v^2} = \frac{m^3 l^2}{v^2}.$$

If we let $v = \sqrt{m^3 l^2}$, $E(|C|) < 1$.

Consider an arbitrary tag not in M. Without loss of generality, suppose it is mapped to the ith segment. In any partition of the segment, the probability for it to share a representative bit with a tag in M is $1 - (1 - k/l)^{m_i}$. The probability for that to happen in all partitions is $[1 - (1 - k/l)^{m_i}]^k$. Hence, the probability for the tag to cause false positive, denoted as p_f is

$$p_f = \sum_{q=0}^{m} Prob\{m_i = q\} \left[1 - \left(1 - \frac{k}{l} \right)^q \right]^k$$

$$< (1 - Prob\{m_i = 0\}) \left[1 - \left(1 - \frac{k}{l} \right)^m \right]^k$$

$$\approx (1 - e^{-lm/v})(1 - e^{-km/l}).$$

The expected value of $|F|$ is

$$E(|F|) = (n - m) \cdot p_f < (n - m)(1 - e^{-lm/v})(1 - e^{-km/l}). \tag{4.14}$$

If we let $v = -ml / \ln[1 - (n - m)^{-1}(1 - e^{-km/l})^{-1}]$ and apply it to (4.14), we have $E(|F|) < 1$. Now, if we choose $v = \max\{\sqrt{m^3 l^2}, -ml / \ln[1 - (n - m)^{-1}(1 - e^{-km/l})^{-1}]\}$, the expected energy cost $E(\Omega) \le E(|C|) + E(|F|) + O(1) < 1 + 1 + O(1) = O(1)$. Therefore, ETOP can also be configured such that the energy cost per tag is $O(1)$.

4.7.2 Execution Time

Following the same analysis as in Sect. 4.5.2, it is easy to see that ETOP has the same formula for protocol execution time as TOP does: $T = (v/l + |F| + |C|)t_{id} + m \times t_{inf} + O(1)$, but the values of $|C|$ and $|F|$ are different. The simulation results in Sect. 4.9 show that ETOP has smaller execution time than TOP.

4.7.3 Choosing v for Time-constrained Energy Minimization

Following the same reasoning in Sect. 4.5.3, we define the time bound for ETOP to be

$$Prob\{T \le B\} \ge \alpha, \tag{4.15}$$

where T is the execution time of ETOP, B is a pre-defined time bound, and α is a probability value, $0 < \alpha < 1$. The objective is to find the largest value v that minimizes the energy cost, subject to the constraint (4.15). In the following, we derive a computable formula for $Prob\{T \le B\}$, which can be found in (4.23) and (4.24). Based on the formula, we will be able to find the optimal value v.

Let m_i be the number of tags in M that are encoded in the ith segment, denoted as V_i, $0 \le i < v/l$. Obviously, m_i follows a binomial distribution $Bino(m, l/v)$,

$$Prob\{m_i = x\} = \binom{m}{x} \left(\frac{l}{v}\right)^x \left(1 - \frac{l}{v}\right)^{m-x}.$$

Let n_i be the number of tags not in M that are mapped to the ith segment, $0 \le n_i \le n - m$. Obviously, n_i follows the binomial distribution $Bino(n - m, l/v)$.

$$Prob\{n_i = z\} = \binom{n-m}{z} \left(\frac{l}{v}\right)^z \left(1 - \frac{l}{v}\right)^{n-m-z}.$$

Let C_i be a subset of C, containing the tags that are removed from V_i; Let F_i be a subset of F, consisting the false positive tags that are mapped to V_i; Let S_i be the union of C_i and F_i, thus $|S_i| = |C_i| + |F_i|$, and,

$$Prob\{|S_i| = s \,|\, m_i = x, n_i = z\} = \sum_{c=0}^{s} Prob\{|F_i| = s - c \,\big|\, |C_i| = c, m_i = x, n_i = z\}$$

$$\cdot Prob\{|C_i| = c \,|\, m_i = x\}. \tag{4.16}$$

Firstly, we show how to calculate $Prob\{|C_i| = c \,|\, m_i = x\}$. After encoding m_i tags in V_i, let d_{ij} be the number of ones in the jth partition, $1 \le j \le k$. As a tag in M has exactly 1 representative bit in each partition, $0 \le d_{ij} \le \min\{m_i, l/k\}$. The reader removes a tag to C_i only if it shares a representative bit with another tag in the partition that contains the largest number of ones. As a result, $|C_i| = m_i - \max_{j \in [1,k]} d_{ij}$. When $y \ge 1$, we have

$$Prob\{\max_{j \in [1,k]} d_{ij} = y \,|\, m_i = x\}$$

$$= \prod_{j=1}^{k} Prob\{d_{ij} \le y \,|\, m_i = x\} - \prod_{j=1}^{k} Prob\{d_{ij} \le y - 1 \,|\, m_i = x\}$$

$$= \left(\sum_{d=0}^{y} Prob\{d_{ij} = d \,|\, m_i = x\}\right)^k - \left(\sum_{d=0}^{y-1} Prob\{d_{ij} = d \,|\, m_i = x\}\right)^k$$

$$= \left(\sum_{d=0}^{y} p_d(x, \frac{l}{k}, d)\right)^k - \left(\sum_{d=0}^{y-1} p_d(x, \frac{l}{k}, d)\right)^k, \tag{4.17}$$

where $p_d(x, l/k, d) = Prob\{d_{ij} = d \,|\, m_i = x\}$ is the conditional probability that a partition containing $m_i = x$ tags happens to have d ones. The calculation of $p_d(\cdot)$ can be found in Sect. 4.5.4. Hence, the conditional distribution of $|C_i|$ is,

$$Prob\{|C_i| = c \,|\, m_i = x\} = \left(\sum_{d=0}^{x-c} p_d(x, \frac{l}{k}, d)\right)^k - \left(\sum_{d=0}^{x-c-1} p_d(x, \frac{l}{k}, d)\right)^k. \tag{4.18}$$

Secondly, we derive $Prob\{|F_i| = s - c \big| |C_i| = c, m_i = x, n_i = z\}$. A tag not in M maps itself to k partitions and choose one bit randomly from each partition. If all these bits are ones, false positive happens. The conditional false positive probability is,

$$Prob\{\text{false positive in } V_i | m_i = x\} = \left(\sum_{d=0}^{x} \frac{kd}{l} p_d \left(x, \frac{l}{k}, d \right) \right)^k . \tag{4.19}$$

When $|C_i| = c$, $\max_{j \in [1,k]} d_{ij} = m_i - c$, hence,

$Prob\{\text{false positive in } V_i \big| |C_i| = c, m_i = x\}$

$$= \frac{1}{p_d(x, \frac{l}{k}, x - c)} \left[\left(\sum_{d=0}^{x-c} \frac{kd}{l} p_d(x, \frac{l}{k}, d) \right)^k - \left(\sum_{d=0}^{x-c-1} \frac{kd}{l} p_d(x, \frac{l}{k}, d) \right)^k \right],$$

$$\tag{4.20}$$

denoted as p_{fc}, which represents the false positive probability when $m_i = x$ tags are encoded in the ith segments and $|C_i| = c$ tags are moved to the collision set C. When n_i tags in $N - M$ are mapped to V_i, the conditional distribution of $|F_i|$ follows the binomial distribution $Bino(n_i, p_{fc})$, thus,

$$Prob\{|F_i| = s - c \big| |C_i| = c, m_i = x, n_i = z\} = \binom{z}{s-c} p_{fc}^{s-c} (1 - p_{fc})^{z-s+c}. \tag{4.21}$$

From (4.18) and (4.21), we can derive $Prob\{|S_i| = s | m_i = x, n_i = z\}$. Thus, the probability distribution of $|S_i|$ is,

$$Prob\{|S_i| = s\} = \sum_{x=0}^{m} \sum_{z=0}^{n-m} \sum_{c=0}^{s} Prob\{|S_i| = s | m_i = x, n_i = z\}$$

$$\cdot Prob\{m_i = x\} \cdot Prob\{n_i = z\}.$$

$$= \sum_{x=0}^{m} \sum_{z=0}^{n-m} \sum_{c=0}^{s} \left[\left(\sum_{d=0}^{x-c} p_d(x, \frac{l}{k}, d) \right)^k - \left(\sum_{d=0}^{x-c-1} p_d(x, \frac{l}{k}, d) \right)^k \right]$$

$$\cdot \binom{z}{s-c} p_{fc}^{s-c} (1 - p_{fc})^{z-s+c} \cdot \binom{m}{x} \left(\frac{l}{v} \right)^x \left(1 - \frac{l}{v} \right)^{m-x}$$

$$\cdot \binom{n-m}{z} \left(\frac{l}{v} \right)^z \left(1 - \frac{l}{v} \right)^{n-m-z}. \tag{4.22}$$

Fig. 4.5 Bound B that satisfies $Prob\{T \leq B\} \geq \alpha$ with respect to v. Parameters: $n = 10,000$, $m = 1,000$.

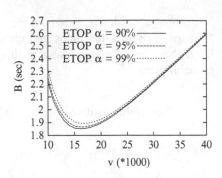

Let S be the union of C and F. We have $|S| = |C| + |F|$ and $|S| = \sum_{i=1}^{v/l} |S_i|$. As S_1, S_2, ..., $S_{v/l}$ are independent of each other, the probability distribution of $|S|$ is the convolution of $|S_i|$. Hence,

$$Prob\{|S| = s\} = Prob\{|S_1| = s\} * ... * Prob\{|S_{\frac{v}{l}}| = s\},$$

where $*$ is the convolution operator. With the help of Fourier Transform, we have

$$Prob\{|S| = s\} = F\hat{F}T\left[\left(FFT\left(Prob\{|S_i| = s\}\right)\right)^{v/l}\right], \qquad (4.23)$$

where FFT is the Fast Fourier Transform, and $F\hat{F}T$ is the inverse Fast Fourier Transform. Adopting (4.7), we have

$$Prob\{T \leq B\} = \sum_{s=0}^{s_{max}} Prob\{|S| = s\}, \qquad (4.24)$$

where $s_{max} = (B - mt_{inf})/t_{id} - v/l$. The right hand side is denoted as $P'_t(v,B)$, which is the probability for the protocol execution time to be bounded by B under a certain value of v. It is computable as a function of v and B after (4.22) is applied and parameters m and n are given. Given a value of B, we can find the largest v that satisfies $P'_t(v,B) < \alpha$ using the False Position algorithm [3]. For example, when $n = 10,000$, $m = 1,000$, $B = 2$ seconds, and $\alpha = 99\%$, the optimal value of v is 23,200.

As a related problem, if v and α are given, we can use $P'_t(v,B)$ to compute the time bound that ETOP can achieve. More specifically, given a value of v, we are able to find the smallest B that satisfies $P'_t(v,B) \geq \alpha$ through bisection search as described in Sect. 4.5.3. Figure 4.5 shows the time bound of ETOP with respect to v when $\alpha = 90\%, 95\%$ and 99%, which correspond to the three curves in the figure.

4.8 Channel Error

Channel error may corrupt the data exchanged between the reader and tags. For example, if a negative polling request is corrupted, the tag that is not supposed to participate in the reporting phase will transmit and cause collision in the reporting phase. A segment of V sent from the reader may be corrupted so that tags encoded in this segment will not report their information. There exists other scenarios of corruption in the execution of TOP or ETOP. They cause two effects: 1) A tag in M does not transmit its information in the slot when it is supposed to transmit, and 2) it transmits but collides with another tag that is not supposed to transmit in the slot. To detect these cases, when a tag transmits, we require it to include a CRC checksum that is computed from the concatenation of the information bits and the tag's ID. When the reader expects information from a tag in a time slot, if the slot turns out to be empty or the data received in the slot do not carry a correct CRC, the reader knows that information from the tag is not correctly received. At the end of the protocol, all missed information can be retrieved by polling the tags directly.

4.9 Simulation Results

We evaluate the performance of the presented protocols, the tag ordering polling protocol (TOP) and the enhanced tag ordering polling protocol (ETOP). We compare them with the basic polling protocol (BP) and the coded polling protocol (CP). The evaluation uses two performance metrics: (1) the average number of bits that each tag has to receive during the protocol execution, and (2) the overall execution time.

We only consider energy consumption of tags in receiving information for two reasons. First, this is the major, variable portion of the energy cost per tag. As we will see shortly, each tag may have to receive hundreds of thousands of bits during protocol execution, whereas it only sends a small, fixed amount, e.g., 32 bits. Second, the energy cost for tags in M to transmit their information is the same for all protocols. Omitting them does not affect the comparison.

We use the following parameters to configure the simulation: each tag ID is 96 bits long, information reported from a tag to the reader is 32 bits long, and each segment in ETOP is 80 bits long and divided into 4 partitions, i.e. $k = 4$. The transmission time is based on the parameters of the Philips I-Code specification [6]. The rate from a tag to the reader is 53 Kb/sec; it takes 18.88 μs for a tag to transmit one bit. Any two consecutive transmissions (from the reader to tags or vice versa) are separated by a waiting time of 302 μs. The value of t_{inf} is calculated as the sum of a waiting time and the time for transmitting the information, which is 18.88 μs multiplied by the length of the information. For 32-bit information, $t_{inf} = 906$ μs. The transmission rate from the reader to tags is 26.5 Kb/sec; it takes 37.76 μs for the reader to transmit one bit. The value of t_{id} is calculated as the sum of a waiting time and the time for transmitting a 96-bit ID. The result is 3,927 μs.

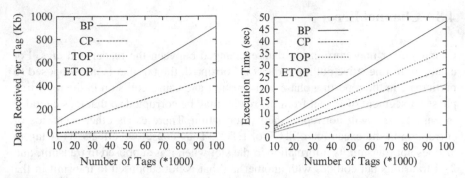

Fig. 4.6 Energy and time comparison. Parameters: $m = 0.1n$, $v = 24m$ for TOP and ETOP. Note that the horizontal '0' line is not at the bottom in order to make the ETOP curve visible.

4.9.1 Varying the Number n of Tags

We first vary the number n of tags in the system from 10,000 to 100,000. We set $v = 24m$ and $m = 0.1n$, i.e., 10% of all tags are selected by the reader to report information. Figure 4.6 compares four protocols in terms of energy cost and protocol execution time. The left plot shows energy costs. TOP and ETOP reduce energy consumption by one or multiple orders of magnitude. For example, when $n = 100,000$, per-tag energy cost in TOP is 9.4% of the cost in CP, and 5.0% of the cost in BP. Per-tag energy cost in ETOP is just 0.52% of the cost in CP, and 0.28% of the cost in BP. The right plot shows the execution time comparison. TOP requires 25% less time than BP, but 27% more time than CP. ETOP requires 55% less time than BP and 24% less time than CP.

In summary, CP reduces both energy cost and execution time nearly by half when comparing with BP. TOP makes great improvement over CP in terms of energy cost, but has modestly higher execution time. ETOP considerably outperforms CP in terms of both energy cost and execution time.

4.9.2 Varying the Size v of Reporting-order Vector

Next, we show how the value of v influences the performance of TOP and ETOP. We set $n = 50,000$ and $m = 5,000, 10,000$, or $25,000$. We vary v from $4m$ to $64m$ and use simulation to find energy cost per tag and protocol execution time. Figure 4.7 shows the simulation results. The top two plots present the average amount of data each tag receives in TOP and ETOP, respectively. The curves match the theoretical results we have given in Sect. 4.5. When v is reasonably large, e.g., $v \geq 7m$, ETOP consumes less energy than TOP. The bottom two plots present the protocol execution time of TOP and ETOP, respectively. ETOP also requires less time than TOP when $v \geq 7m$.

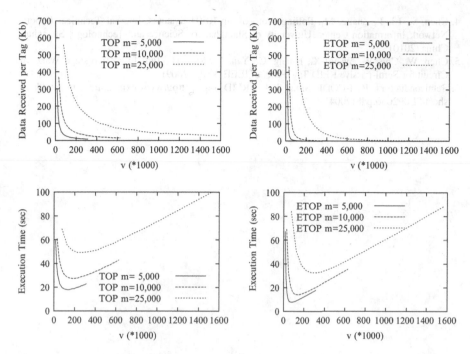

Fig. 4.7 Energy cost and execution time comparison of TOP and ETOP. **First plot (top left)**: Energy cost of TOP with respect to v. **Second plot (top right)**: Energy cost of ETOP with respect to v. **Third plot (bottom left)**: Execution time of TOP with respect to v. **Fourth plot (bottom right)**: Execution time of ETOP with respect to v.

4.10 Summary

In this chapter, we present two energy-efficient polling protocols, TOP and ETOP, for large-scale RFID systems. These protocols are designed to collect real-time information from a subset of tags in the system. Our primary objective is to lower energy consumption by tags in order to extend their battery lifetime. The new protocols can be configured to achieve $O(1)$ energy cost per tag. Performance tradeoff between energy cost and execution time can be made by controlling the size of the reporting-order vector.

References

1. Bloom, B.: Space / Time Trade-offs in Hash Coding with Allowable Errors. Communications of the ACM **13**(7), 422–426 (1970)
2. Broder, A., Mitzenmacher, M.: Network Applications of Bloom Filters: A Survey. Internet Mathematics **1**(4), 485–509 (2004)
3. Burden, R., Faires, J.: Numerical Analysis (2011)

4. Chen, S., Li, J., Zhang, M.: Polling Protocols in RFID Systems. Internal Technical Report,
   Network Information Center, University of Shanghai for Science and Technology, Shanghai,
   China (2010)
5. Chen, W., Che, W., Wang, X., Huang, C., Yan, N., Min, H., Tan, J.: A Two-stage Wake-up
   Circuit for Semi-passive RFID Tag. Proc. of IEEE ASIC (2009)
6. Semiconductors, P.: I-CODE Smart Label RFID Tags. http://www.nxp.com/documents/data_
   sheet/SL092030.pdf (2004)